IC VOLTAGE REGULATOR SOURCEBOOK WITH EXPERIMENTS

IC VOLTAGE REGULATOR SOURCEBOOK WITH EXPERIMENTS

BY VAUGHN MARTIN

TAB BOOKS Inc.
BLUE RIDGE SUMMIT, PA. 17214

FIRST EDITION

FIRST PRINTING

Printed in the United States of America

Library of Congress Cataloging in Publication Data

Martin, Vaughn, 1946-
IC voltage regulator sourcebook, with experiments.

Includes index.
1. Voltage regulators. 2. Integrated circuits.
I. Title. II. Title: I.C. voltage regulator sourcebook, with experiments.
TK7872.V55M37 1983 621.3811'73 83-4858
ISBN 0-8306-0157-0
ISBN 0-8306-0557-6 (pbk.)

Contents

Introduction

Like the Roman God Janus, who faces both ways, the *IC Voltage Regulator Sourcebook, with Experiments* addresses the needs of the beginner and hobbyist as well as the serious experimenter, student, technician, and practicing engineer by providing regulator and power supply circuitry with formulas, which make modification possible. The text provides an explanation of each circuit and also explains the circuit's pitfalls. As an example, latch-up in foldback current limiting designs is examined and alternative circuits are provided to help avoid these kinds of problems.

Basic theory is provided throughout. The Bibliography provides references to enable the more industrious to further investigate particular subjects of interest. Experiments at the end of the first five chapters are designed to reinforce the principles covered in those chapters. This book also explains semiconductor manufacturers' parts numbering conventions for voltage regulator ICs.

The basics of what is inside a modern voltage regulator IC is covered early in Chapter 2 so that you can better understand and more fully appreciate the discussions on newer technological advances in voltage regulators covered later. The book is primarily directed toward the hands-on type of reader, who requires a quick working design to satisfy some particular circuit need. As examples, there are tables listing the output voltage and current of various regulators and other tables list numerous other comparative features of both linear and switching voltage regulator ICs. There is an appendix with voltage regulator data sheets provided for reference.

The first three chapters cover power supply and regulator basics; Chapter 4 examines switching voltage regulators by looking at three of the most commonly used switching voltage regulator ICs in use today. Tabularized formulas are provided for some of these regulators.

The 723 voltage regulator IC chapter is included, not as a representative of a modern IC voltage regulator, but rather as an example of a multiterminal precision voltage regulator IC. This type of regulator has been around for about ten years and has had so many designs generated around it that its exclusion would be inappropriate.

The last three chapters further distinguish this from any other voltage regulator book written thus far. Chapter 6 deals with virtually every critical power supply parameter imaginable that might be monitored—including power-on risetime, thermal shutdown time, power shutdown delay from either overvoltage or current, etc. Additionally, simple voltage threshold detecting circuits and numerous visual continuously monitoring circuits are included. Chapter 7 examines methods to maximize power utilization and efficiency in power supply and voltage regulator designs. Eight methods are listed and examined separately.

The last chapter provides user guidelines and design procedures for heatsinks and discusses fans, blowers, and thermoelectric devices such as Peltier effect devices which pull heat away from semiconductor surfaces. There is also an explanation of passive temperature indicators that change color in response to varying degrees of heat exposure and overheating protection circuits using PTCs (positive temperature coefficient thermistors).

The following individuals deserve thanks for providing me with illustrations, permission to reprint numerous circuits, and samples of various voltage regulator ICs:

Bill Brennan	National Semiconductor
Clive McCarthy	Fairchild Semiconductor
Peter Bradshaw	Intersil
Murry Shohat	Signetics
Stan Dendinger	Silicon General
Thomas A. Skornia	Advanced Micro Devices

Lastly, I want to thank Kathy Covert for typing this manuscript and Paula Anderson and my mother for their support throughout this effort.

IC VOLTAGE REGULATOR SOURCEBOOK WITH EXPERIMENTS

Chapter 1

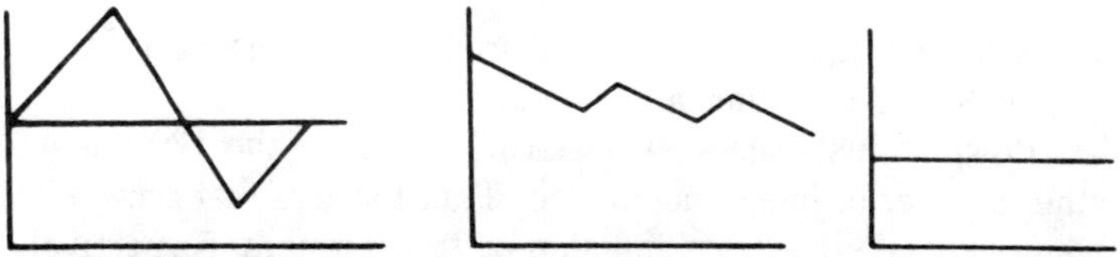

Transformers, Rectifiers, and Filters

A dc (direct current) power supply can be designed with only a transformer, one to four diodes, a capacitor, and a solid-state IC (integrated circuit) voltage regulator, or it may require many more active (ICs and transistors) and passive (resistors, capacitors, and inductors) components. The capabilities of these two approaches or levels of design sophistication are naturally quite different. However, no matter how the design is approached, the basic elements shown in Fig. 1-1 must be present.

The transformer converts or transforms the 120 Vac (volts alternating current from the power company) to a lower voltage somewhere between 6 Vac to 40 Vac. The rectifier section, consisting of one to four diodes, rectifies or changes the ac into pulsating dc, which the filter capacitor smooths. Figure 1-2 illustrates this filtering or smoothing action. The filtering section may consist of just a capacitor, a capacitor and an inductor, or a capacitor and resistor configuration. (See Fig. 1-3.) This chapter discusses the transformer, rectifier and filter sections of a power supply.

TRANSFORMERS

Transformers consist of two coils placed closely together. They are usually built by wrapping two sets of wires around a piece of iron called a core. We are interested solely in power transformers operating at 60 Hz (cycles per second) and not in air core or small impedance matching coupling transformers that are very small and sometimes collectively called subouncers. These transformers

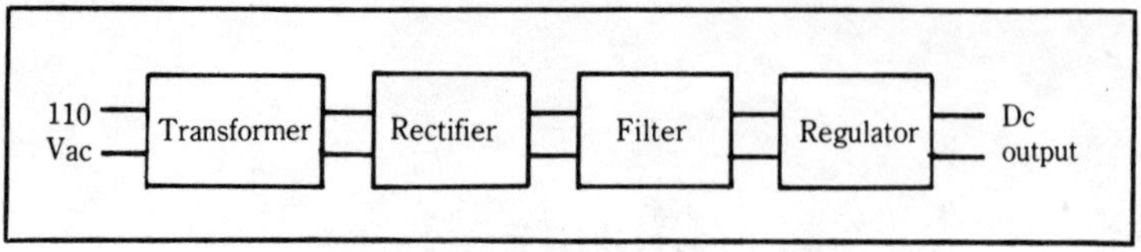

Fig. 1-1. Dc power supply block diagram.

couple audio frequency signals and are found in stereos, TVs, tape recorders, and communications equipment.

Power transformers "transform" or change the 120 Vac into either a lower or higher ac voltage. Transformers that accomplish this are called step-down and step-up transformers, respectively. The amount of voltage being stepped up or down is in direct proportion to the turns ratio. For example, if the primary has 12 times the number of turns the secondary has, then the secondary voltage will be 1/12 or 10 Vac with a 120 Vac input on the primary.

There are two different secondary winding configurations. One simply has two wires on the output (secondary) and the other has three wires on the secondary. This extra wire is a tap right in the center of the secondary winding. This center tap allows a convenient spot to tap off exactly one-half of the full ac potential from the secondary. This allows full-wave rectification, which will be discussed shortly. Some transformers (usually the larger ones) have multiple secondaries as shown in Fig. 1-4. The most common single center-tapped secondary has two red leads and a yellow lead. The yellow is the center-tap lead. The two black leads on the other side are the primary.

Note the dots at the top of the primary and secondary in Figs. 1-4 and 1-5. This is a common convention indicating that the input and output are in phase with each other. If the secondary in Fig. 1-5 had a dot at its bottom, this would indicate that the input and output were exactly 180 degrees out-of-phase. The multiple-winding

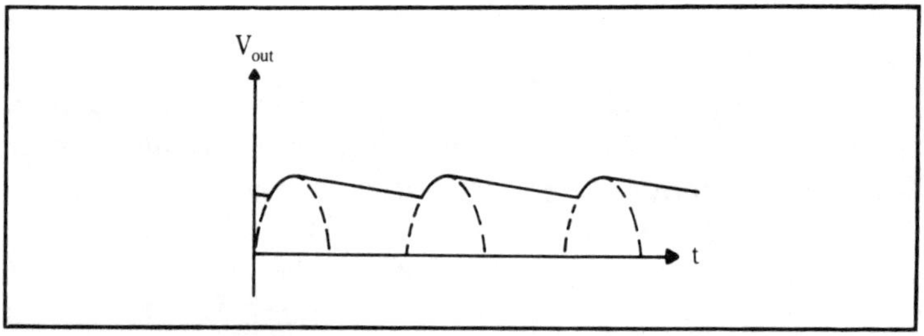

Fig. 1-2. A capacitor's filtering action.

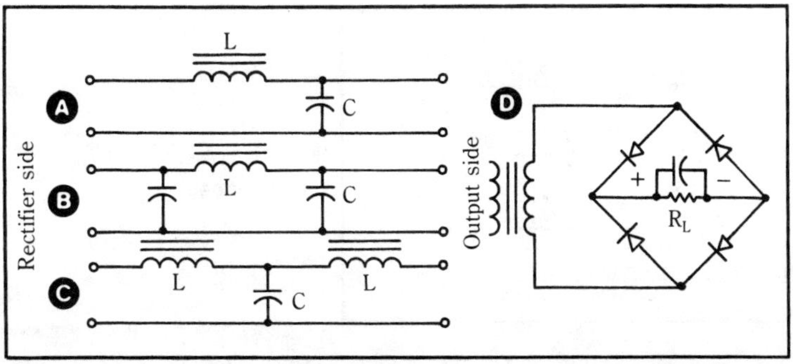

Fig. 1-3. Filtering section configurations. Filters using both an inductor and a capacitor: (A) L-section filter; (B) π-section filter; (C) T-section filter; (D) RC-section filter.

transformers may also be "stacked" or have their secondaries connected together to give an additive effect. For example, if the multiple-winding secondary has two 12-volt secondary windings, and if the top of the bottom winding with the dot is connected to the bottom of the top winding (Fig. 1-4), the two voltages add or 24 V results. Do not connect windings with the dots together. It may be easiest to visualize these dots as positive terminals on a battery and naturally batteries are connected together in series for an additive voltage by the adjacent positive to negative terminal connections. The same is true for transformer secondaries.

Figure 1-4 is an illustration of a multiple-secondary transformer with leads color coded. Note the tap on the primary. If a secondary voltage of approximately 93 percent of the values present with the black to black-red primary is desired, the black-yellow

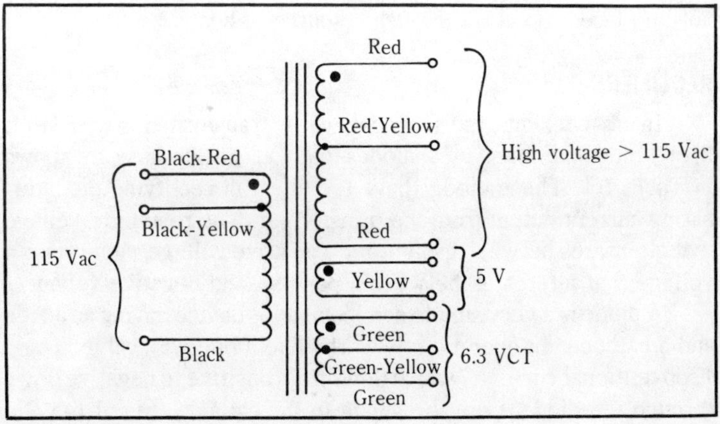

Fig. 1-4. Multiple-secondary transformers.

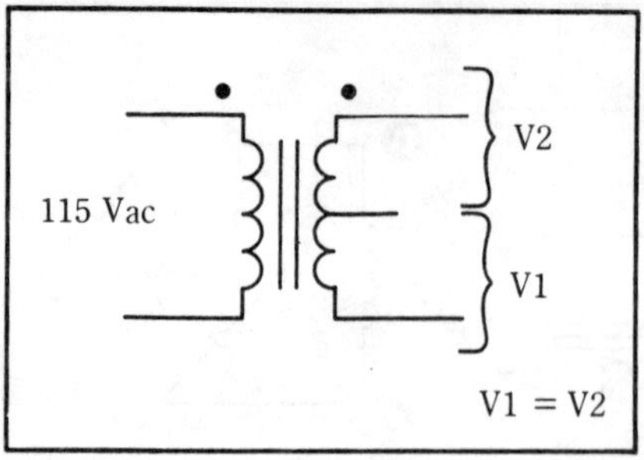

Fig. 1-5. A center-tapped transformer.

lead may be used instead of the black-red lead. These colors are an EIA (Electronic Industries Association) standard.

Transformers are specified for a certain output voltage at their full current rating. If less than this full current is drawn, the output voltage will be greater than the specified output voltage. Usually with no load the output voltage is 20 percent greater than specified and decreases as more current is drawn until maximum current drawn is reached, at which point the output voltage is as specified. As the transformer becomes larger the regulation improves from this 20 percent figure. Transformers are also not 100 percent efficient, that is, the product of the input current times voltage is greater than the product of the output voltage times current. This "inefficiency" is due to the resistance of the wire in the windings and core losses. The more iron used in the transformer the less the core losses, but this can quickly become a disadvantage in cost and weight. The power lost in a transformer takes the form of internally generated heat. When operating a transformer at maximum ratings it can become hot to touch and heat-sensitive analog circuitry should not be placed too close to such a source of heat.

RECTIFIERS

In most designs, the secondary of the transformer is wired to 1, 2, or 4 diodes. Some typical diodes that you can purchase are shown in Table 1-1. These diodes have the effect of rectifying the alternating current output from the transformer. The result is a voltage that alternates between ground and a positive voltage, rather than a voltage that alternates between a positive and negative voltage.

A diode is a very simple semiconductor device having an anode and a cathode. In operation, it works like a one-way valve. Thus, "conventional current," which flows from positive to negative flows through the diode, from the anode to the cathode (Fig. 1-6). The

Table 1-1. Specifications for the 1N4000 Series Diodes.

MAXIMUM RATINGS Rating	Symbol	1N4001	1N4002	1N4003	1N4004	1N4005	1N4006	1N4007	Unit
Peak repetitive reverse voltage or working peak reverse voltage or dc blocking voltage	V_{RRM} V_{RWM} V_R	50	100	200	400	600	800	1000	Volts
Nonrepetitive peak reverse voltage (halfwave, single phase, 60 Hz)	V_{RSM}	60	120	240	480	720	1000	1200	Volts
RMS reverse voltage	$V_{R(RMS)}$	35	70	140	280	420	560	700	Volts
Average rectified forward current (single phase, resistive load, 60 Hz, $T_A = 75°$ C)	I_0	1.0	1.0	1.0	1.0	1.0	2.5	2.5	Amp
Nonrepetitive peak surge current (surge applied at rated load conditions)	I_{FSM}				30 (for 1 cycle)				Amp
Operating and storage junction temperature range	T_J, T_{stg}				−65 to +175				°C

diode blocks "electron current" flowing from the cathode to the anode.

Since diodes are semiconductors (they don't conduct as well as a copper wire), there is a voltage drop across them because of their internal resistance. This resistance is usually fairly small, so you can expect a voltage drop of only about 0.6-0.7 V across the diode. Thus, you might measure 11.8 V on the anode, and 11.2 V on the cathode. When semiconductor diodes were first made, they were made of germanium. Germanium diodes, because of their lower internal resistance, only had voltage drops of 0.3 V. However, now silicon diodes are much more common, and they have the 0.6-0.7 V voltage drop that we just mentioned. Like transformers, diodes

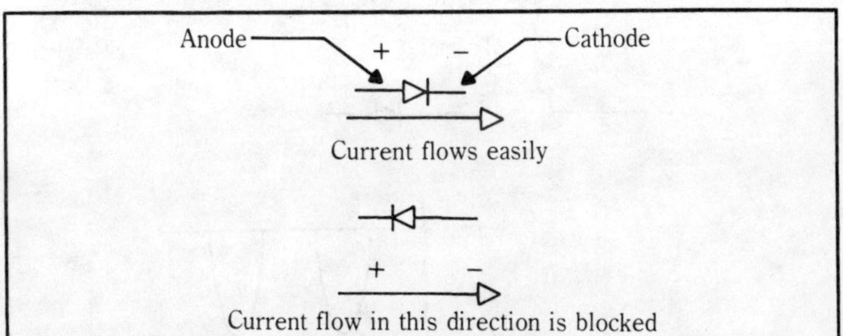

Fig. 1-6. Current flow through a diode.

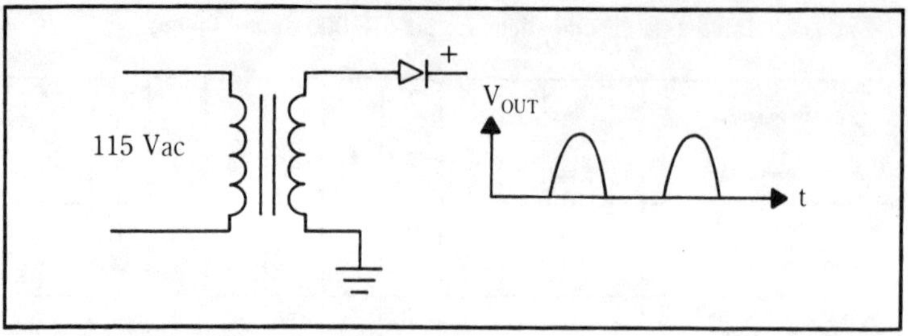

Fig. 1-7. A half-wave rectifier (positive output).

dissipate heat when current flows through them. The more current, the hotter they get. You can specify the current-handling capabilities (1 A, 2 A, 5 A, etc.), and you will have to consider this when you start designing your power supply. In fact, in some power supply designs, the diodes have to be mounted on a heatsink!

Diodes are not perfect devices; therefore, it is possible to put a high enough positive voltage between the cathode and anode so that a current will flow in the reverse direction. This is called the *peak inverse voltage* (Table 1-1). The better the diode, the higher this voltage will be, and the more expensive it will be. In your designs, you must not exceed the voltage as given in Table 1-1.

The simplest rectifier configuration is the half-wave rectifier. As its name implies, this configuration (which uses only one diode), only allows half of the ac waveform through (Fig. 1-7). Because only one-half of the waveform passes through, only the positive portion of the waveform is present on the diode's output. The diode can also be reversed (Fig. 1-8) so that only the negative portion of the ac is present on the anode of the diode.

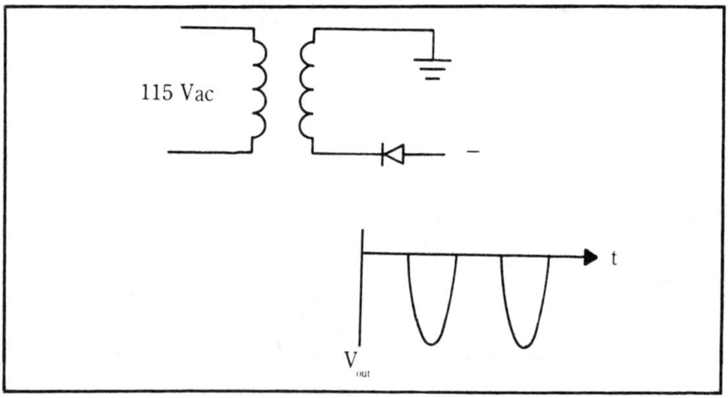

Fig. 1-8. A half-wave rectifier (negative output).

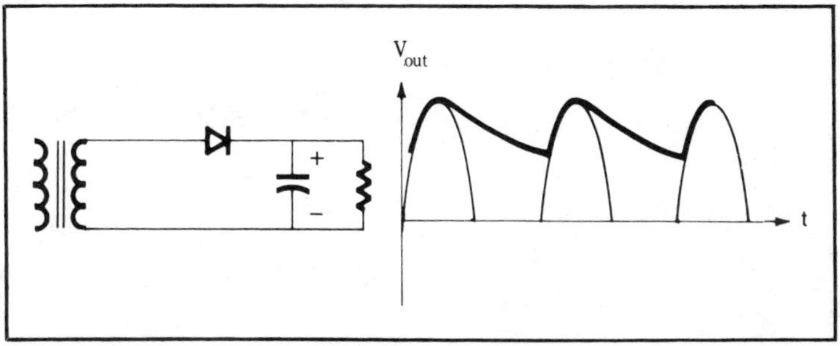

Fig. 1-9. A half-wave rectifier output waveform.

This rectifier configuration is simple, requiring no transformer center-tap and only one diode, but it has some disadvantages that limit its use to power supplies of one watt or less. This configuration is harder to filter because of the alternate half cycles that go to zero volts and allow the filter capacitor to "droop" (Refer to Fig. 1-9). Also, the current flows in just one direction in the transformer secondary so the core may saturate. This means that if large currents are drawn through the secondary the magnetic field in the iron doesn't "collapse" or disappear when the voltage on the primary goes to zero. The result is that for a given amount of secondary current, a half-wave rectifier will require a larger, more costly transformer with more iron in its core for performance comparable to another rectifier configuration.

A much more common and efficient rectifier design is the full-wave rectifier (Fig. 1-10). Note that the full-wave rectifier uses a center-tapped transformer, which is sometimes hard to obtain in the proper voltage. Also, there are no "bare spots" in the output of this rectifier, like there were with the half-wave rectifier. The frequency of the output voltage is also twice that of the input voltage. Thus, if the alternating current in the primary has a frequency of 60 Hz, the output will have a frequency of 120 Hz.

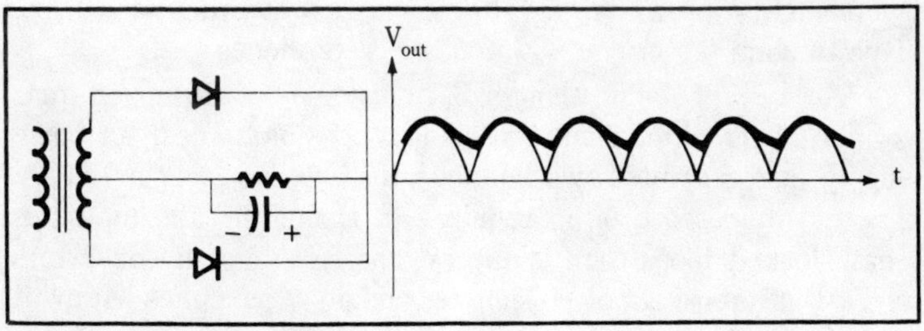

Fig. 1-10. A full-wave rectifier.

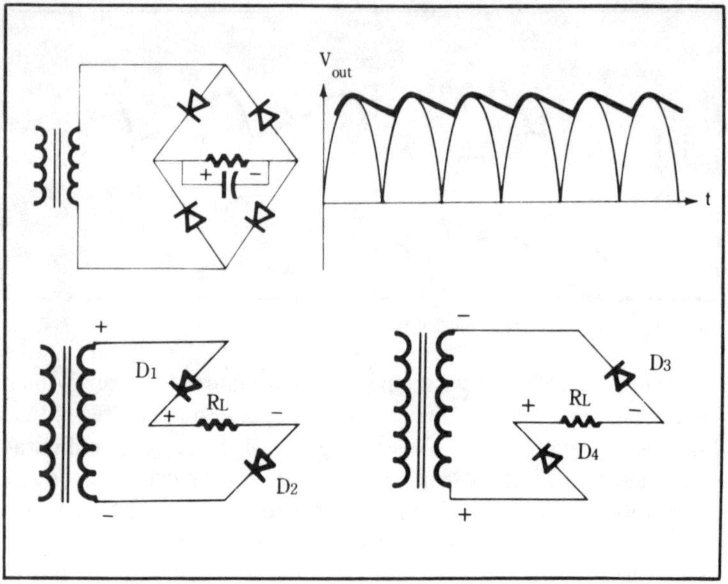

Fig. 1-11. A bridge rectifier. The two different paths through which current flows during alternate half-cycles.

Another common rectifier design is the bridge circuit shown in Fig. 1-11. In this circuit, a center-tapped transformer is not required. Another advantage that the bridge rectifier has over the full-wave rectifier is that the inverse voltage applied to the bridge diodes is one-half of that applied to the full-wave diodes. This is because the voltage is "dropped" across two diodes in the bridge circuit, while it is only dropped across one diode in the full-wave rectifier.

It isn't surprising that many power supplies are built using diode bridges, since they don't require a center-tapped transformer. The semiconductor manufacturers have also come to our aid, because you can actually purchase a diode bridge with a specified PIV (peak inverse voltage) and current rating, so you don't have to build the bridge with four discrete diodes.

As we will see in Chapter 3, by using a center-tapped transformer, along with a diode bridge, it is possible to design a power supply that generates two voltages, such as ±5, ±12 or ±15 V. These voltages are often required in analog circuits (op amps, phase-locked loops, active filters, analog-to-digital converters, etc.). Dual-tracking power supplies and power supplies with voltages of equal magnitude but opposite polarities require a center-tapped transformer to generate a positive and negative output voltage.

FILTERS

One problem that we still have with the full-wave and bridge rectifiers is that the output of the rectifier still goes to zero volts, even though it is at zero volts for a lot less time than the half-wave rectifier. If we want to power a microprocessor, tape recorder, digital logic or radio, the voltage must not drop to zero. To prevent this, a capacitor is wired across the outputs of the rectifier.

When this is done, the capacitor acts like a storage device. When the output of the rectifier is at its peak voltage, the load draws current through the transformer and rectifier. Once the voltage starts to drop, and the voltage on the primary goes towards zero, the load starts drawing charge (electrons) from the capacitor. If the capacitor is large enough, the load will need relatively few electrons from the capacitor, so the voltage on the capacitor varies very little. If the capacitor is small, the load will draw so many electrons from the capacitor that the voltage will drop half or more, towards zero volts.

This change in voltage is called *ripple* and is expressed in volts. If the peak voltage on the rectifier's or capacitor's output is 12 V, and the capacitor is discharged to 11.5 V when the primary voltage drops to zero, the ripple would be 0.5 V. Ripple is also sometimes expressed as a ripple factor. For the power supply whose output drops from 12 V to 11.5 V, the ripple factor would be 0.0434 (0.5/11.5). The size of the capacitor that you will need in your power supply will be determined by (1) the frequency of the waveform on the rectifier's output (either 60 Hz or 120 Hz in most of our designs), (2) the amount that the voltage output of the capacitor can drop, and (3) the amount of current required by the load. The effect that these three parameters have on the capacitance is summarized in Table 1-2.

Table 1-2. The Effect of Frequency, Current, and Ripple on the Size of the Filter Capacitor.

Constant	Constant	Change	Capacitance
Ripple	Current	Increase Frequency	Decrease
Current	Frequency	Increase Ripple	Decrease
Frequency	Ripple	Increase Current	Increase
Ripple	Current	Decrease Frequency	Increase
Current	Frequency	Decrease Ripple	Increase
Frequency	Ripple	Decrease Current	Decrease

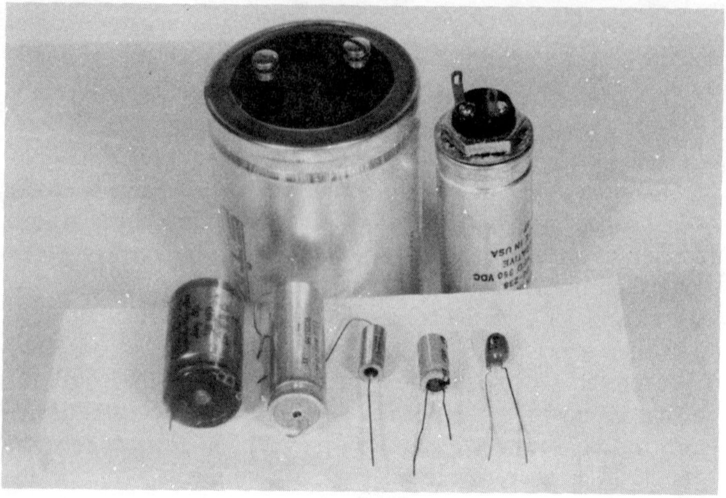

Fig. 1-12. Capacitors in varying sizes and shapes.

Mathematically, this relationship can be expressed as:

$$C = \frac{I}{dV} \times 0.006 \qquad \textbf{Equation 1-1}$$

where I = amount of current required by the load (amps)
dV = ripple (in volts)
C = capacitance (in farads)

For example, if 0.5 V peak-to-peak ripple is acceptable in a 1 A power supply, the required capacitance would be:

$$C = \frac{1\ A}{0.5\ V} \times 6 \times 10^{-3} = 12{,}000\ \mu F$$

This result is based on the assumption that the frequency of the ripple is 120 Hz, which would be true if a full-wave or bridge rectifier is being used.

Capacitors vary in size and shape (Fig. 1-12). The larger capacitors are polarized and are called electrolytics. These have a + and – terminal with the more positive potential being applied to the + terminal. If care is not exercised and these polarities are reversed, capacitors may explode and blow out a gas plug from their ends. The most common type electrolytic capacitors are aluminum electrolytics. These are made by having foil "rolled" or coiled up in an overlapping concentric fashion with a paper or other insulator serving as the "dielectric" which separates the adjacent conductors from shorting one another. There are techniques of etching this aluminum to increase its effective surface area and in turn its capacitance. Aluminum electrolytics suffer inductance from the coiling of their conductors. Tantalum electrolytics are constructed differently, resulting in less inductance. Tantalum itself is a rare element that is corrosion-resistant. These capacitors are more expensive but have superior noise-filtering capabilities over aluminum electrolytics. Most electrolytics have voltage ratings expressed as WVdc (working volts dc). If this voltage rating is exceeded, the insulator or dielectric will break down and the two conductors will short together. This can result in either a decreased capacitance rating or a destroyed capacitor that appears as a complete short.

INDUCTOR-CAPACITOR AND RESISTOR-CAPACITOR SECTIONS

It is also possible to filter the rectifier's output using LC and RC networks (Fig. 1-13). Any number of these sections can be put in series in order to decrease the amount of ripple. However, one problem with the inductor-capacitor sections is that inductors that have a large inductance and at the same time can pass 1 A or 2 A of current are bulky and expensive. It's a lot easier and cheaper to increase the size of the capacitor simply by placing two or three in parallel if you don't have a really big one. You can now buy a relatively small 2-3 in. diameter, 1μF capacitor. The same goes for

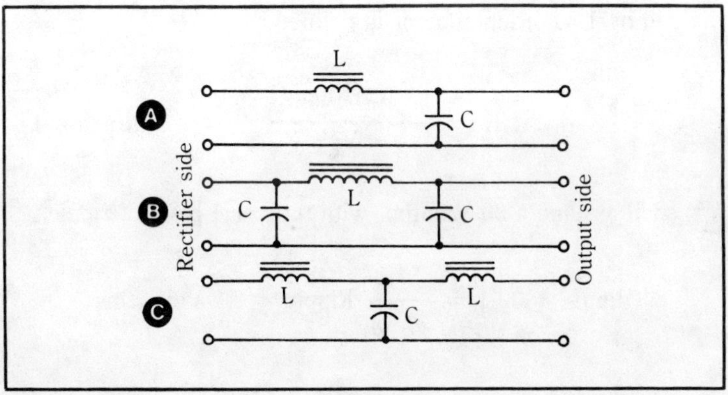

Fig. 1-13. LC filtering sections. (A) L-section filter; (B) π-section filter; (C) T-section filter.

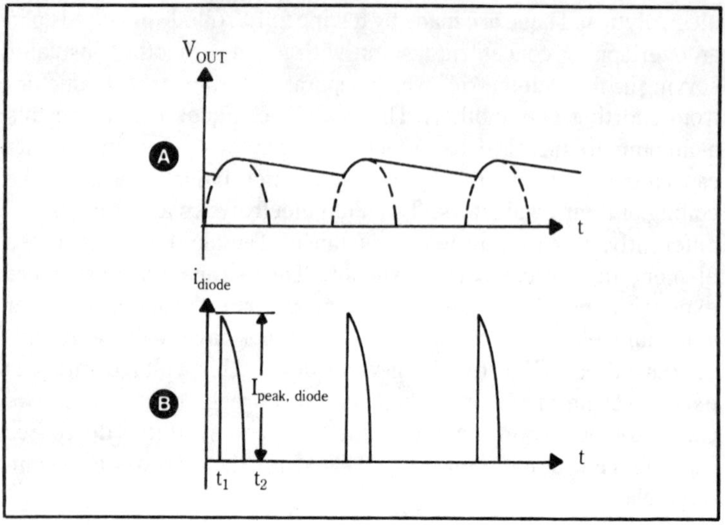

Fig. 1-14. Peak surge current in a diode. (A) Output voltage; (B) diode current.

using RC sections, although high-power low-resistance resistors are more common than high-inductance, high-power inductors.

CALCULATING COMPONENT VALUES

Now that we have studied each part of a basic power supply with the exception of the voltage regulator, let's go through a design problem. This design is for a full-wave rectified sine wave output. First, let us examine ripple. As previously stated, ripple is an ac voltage that is not entirely filtered or smoothed out. This is expressed in volts rms (root-mean-square) which is the peak voltage divided by 1.41. Ripple factor is defined as:

$$K_r = \frac{V_{rms} \text{ Ripple}}{V_{dc}} \qquad \textbf{Equation 1-2}$$

We shall assume a 20 V output with 0.01 or 1 percent ripple.

$$\text{Ripple} = 0.01 = \frac{V_{rms}}{20\text{ V}} \quad \text{Ripple} = 0.2 \text{ Volts rms}$$

Suppose our requirements are a I_{dc} load = 500 mA and a V_{dc} load = 20 V.

$$\begin{aligned} PIV &= 2\,(V_{dc}\text{ out} + \text{V Peak Ripple}) \\ &= 2\,(20 + 2\,(.2)) \qquad \textbf{Equation 1-3} \\ &= 40.8\text{ V.} \end{aligned}$$

$$\begin{aligned} \text{where } I_{Av}\text{ per diode} &= (1/2)\,(I_{Av}\text{ Out}) \\ &= (1/2)\,(500\text{ mA}) \qquad \textbf{Equation 1-4} \\ &= 250\text{ mA} \end{aligned}$$

$$I_{\text{Recurrent Peak}} \simeq I_{\text{dc Load}}\left(1 + 2\,\pi\sqrt{\frac{1}{\sqrt{2}\,K_r}}\right) \qquad \textbf{Equation 1-5}$$

$$\simeq 500\text{ mA}\left(1 + 6.24\left(\sqrt{\frac{.707}{.01}}\right)\right)$$

$$\simeq 500\text{ mA}\,(1 + 6.24\,(8.41))$$

$$\simeq 500\text{ mA}\,(1 + 52.48) = .5\,(53.48)$$

$$\simeq 26.24\text{ Amps surge}$$

where K_r = ripple factor

Recurrent peak current is also probably more commonly known as surge current and will appear as such on diode data sheets. This magnitude of surge current 26.24 A, is probably a bit surprising to you; however, consider carefully what is occurring in a filtered power supply. By the law of conservation of electric charge, it is apparent that the amount of electric charge sent to the capacitor during the conduction portion of the cycle must equal the amount of charge feeding the load over the rest of the cycle. This may make more sense when you take another look at the peak recurrent current Equation 1-5 for the diodes. The current is inversely proportional to the ripple factor, or the less ripple the more the peak current. Refer to Fig. 1-14 for an illustration of this effect. Note how little time there is for this charge to occur, thus, the resultant surge current is understandably quite high.

We have already expressed a relationship for determining the minimum-sized capacitor needed when ripple and load current are known. The following equation approximates the minimum capacitance required when ripple factor and load resistance are the two quantities to be solved for:

$$C \simeq \frac{1}{2\sqrt{2}\,K_r\,R\,f} \qquad \textbf{Equation 1-6}$$

where f = 60 Hz for halfwave, 120 Hz for full-wave
K_r = ripple factor
R = load resistance in ohms

Returning to our previous design example, we were delivering 20 V at 0.5 A so the effective load by Ohm's law is 40 ohms. The ripple factor desired was 0.01 or one percent and the frequency (f) is 120 Hz with any non-half-wave rectifier configuration. Plugging these numbers into Equation 1-6 yields a 7,366 microfarad capacitor. Actually, 10,000 microfarads at 35 WVdc would be a wise choice. This is a standard size and the added capacitance will make the ripple diminish by approximately 35 percent. Additionally, capacitors are notoriously inaccurate in their values. A typical capacitor may be 50 percent over or 10 to 20 percent below its specified capacitance value.

EXPERIMENT 1-1.

This experiment shows that a full-wave rectifier has less output ripple than a half-wave rectifier.

1. Select one electrolytic capacitor in the 10 to 100 microfarad range.
2. Construct the half-wave rectifier circuit shown in Fig. 1-9, carefully noting capacitor polarity.
3. Measure the peak-to-peak ripple on the output with an oscilloscope.
4. With the same capacitor and load resistor as used in step 2, construct the full-wave rectifier shown in Fig. 1-10.
5. Measure the peak-to-peak ripple on the output with an oscilloscope and compare this to the results obtained in step 3. This value should be smaller than the results obtained in step 3.

Chapter 2

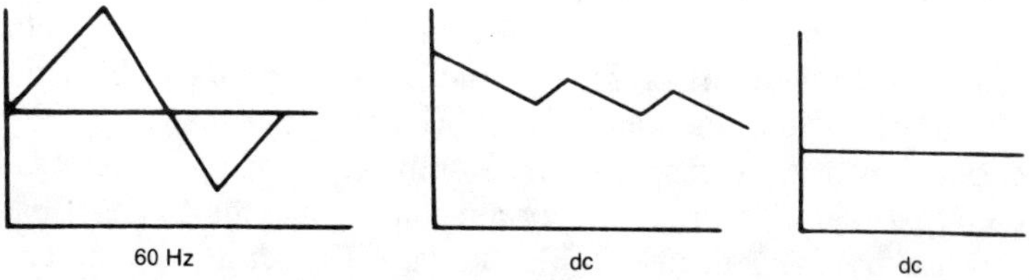

The Basic Voltage Regulator

Up until the early 1970s, if you wanted to build a well-regulated power supply, you had to know a lot about transistors and be comfortable designing a fairly complex circuit. A simple power supply could be designed with just a zener diode, a power transistor, and a resistor or two, but a well-regulated power supply whose output didn't vary with increases or decreases in load current or temperature was a complex undertaking. All this changed with the introduction of the modern integrated circuit voltage regulator. Three of the earlier voltage regulators were the LM100, LM723, and LM309K. The LM100 is a variable positive regulator, the LM309K is a fixed 5 V, 1 A regulator, and the LM723 precision voltage regulator is a very versatile, positive or negative power supply building block. This particular regulator is so versatile and popular that Chapter 5 is devoted entirely to it.

The introduction of these regulators along with others has greatly simplified power supply designs. As an example, when power supply designs revolved around discrete transistors, ripple rejection for the regulator was a problem. Now with integrated circuits, a ripple rejection of 60 dB is not uncommon. What this means is that the output varies no more than 100 μV when the input varies 100 mV, a (1,000:1) ratio. This problem is so trivial now that many voltage regulator data sheets do not even list this parameter.

There are a number of fixed-output voltage regulators on the market today, mostly three-terminal devices with either positive or negative outputs in the 5, 6, 12, 15, and 24 V range. The three-terminal IC voltage regulator pins are as follows: an input for an unregulated input voltage, a ground or adjustment reference terminal, and an output terminal that supplies the specified regulated output voltage.

There are numerous manufacturers of integrated circuit voltage regulators and all seem to have their own parts numbering system. There is one redeeming point to what might appear to be utter chaos—usually only the prefixes and suffixes to a part number change. The basic part number like the 723, 309, 7800 series etc. remain intact with the prefixes mainly indicating the specific manufacturer. National uses an LM and LH prefix for linear monolithic or linear hybrid ICs, respectively. Motorola uses an LML prefix and Fairchild uses a μA prefix. Many components like the 109, 209, and 309 are basically identical. The initial "1" indicates a military temperature range and generally much tighter specs. The "2" in the basic number indicates an industrial grade component with an initial "3" indicating a commercial temperature range IC.

There are also families of integrated circuit voltage regulators with numbers indicating many key characteristics. As an example, the 7700 series is a series of 750 mA voltage regulators with positive outputs. A 7705 would be just such a regulator with a 5 V output. The 7800 series has 1.5 A output current in a wide variety of voltages and the 78L00 is a small package, 100 mA, fixed regulator.

Table 2-1. A Selection Guide for Some Adjustable and Fixed-Voltage Regulators.

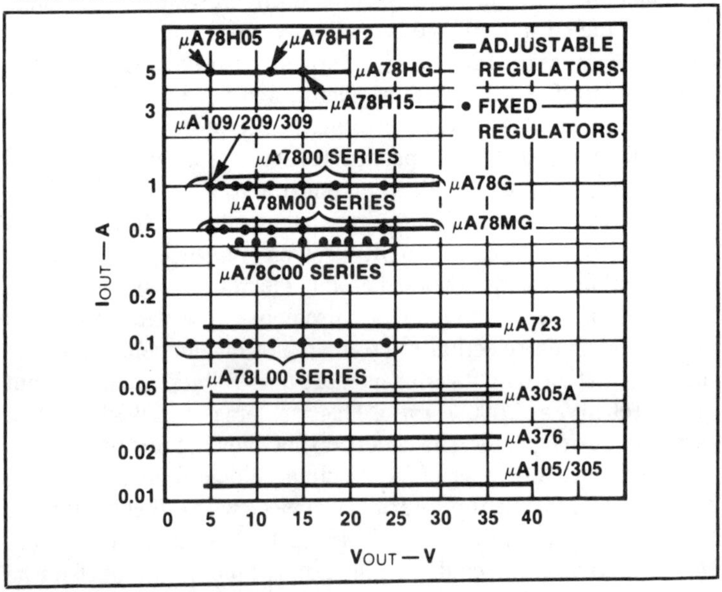

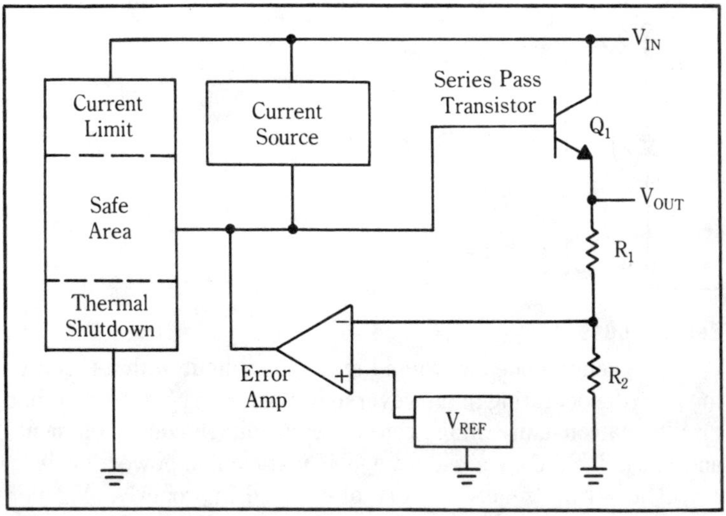

Fig. 2-1. Regulator block diagram. (From *National Semiconductor Voltage Regulator Handbook,* 1980 National Semiconductor; reprinted with permission.)

The 78M00 has 0.5 A output and the 7900 series is identical to the 7800 series with the exception that they have a negative output. The 79L00 is a negative regulator with 100 mA output. All of these regulators come in a wide variety of output voltages and with both positive and negative versions it is easy to construct a ± or bipolar power supply from them. Refer to Table 2-1 for a comparison of regulators.

Despite this variety, the internal circuitry of most voltage regulators is essentially the same. A voltage regulator block diagram is shown in Fig. 2-1. There are five elements that make a voltage regulator. These are a voltage reference, a constant-current source, an error amplifier, a series-pass transistor, and some protection circuitry.

THE VOLTAGE REFERENCE

A voltage reference may consist of a zener diode, a bandgap reference, or a zener diode and transistor. Ideally, the voltage reference's voltage should not drift or change with time, temperature, or varying output load requirements. This reference voltage is the most crucial element in the voltage regulator because any errors developed in this source will be amplified, along with the reference voltage, by the error amplifier. Zener diodes are one of the most popular voltage references.

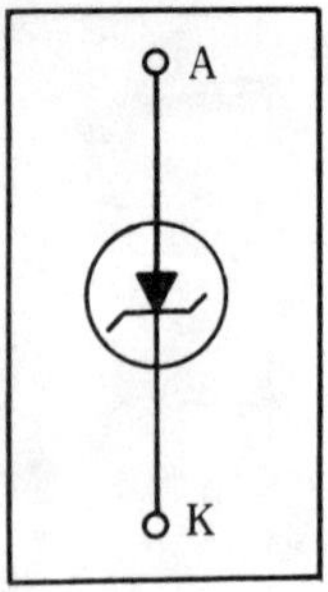

Fig. 2-2. The zener diode.

Zener Diodes

The zener diode, shown in Fig. 2-2, is a diode with its rectifying junction operating in the reverse breakdown region. The zener provides a constant voltage reference or voltage control element, and is available over a wide range of voltage and power levels.

The term "zener" is very often used improperly. A zener diode is a diode with a *negative temperature coefficient.* That is, the value of the breakdown voltage (the zener voltage) decreases as the PN junction temperature increases. An avalanche diode's breakdown voltage increases as the PN junction temperature increases; therefore, the avalanche diode has a *positive temperature coefficient* (Fig. 2-3). Diodes with a breakdown voltage of less than 5 volts exhibit predominately zener characteristics with respect to temperature. Both zener and avalanche characteristics are present in diodes with breakdown voltages between 5 and 8 volts. Above 8

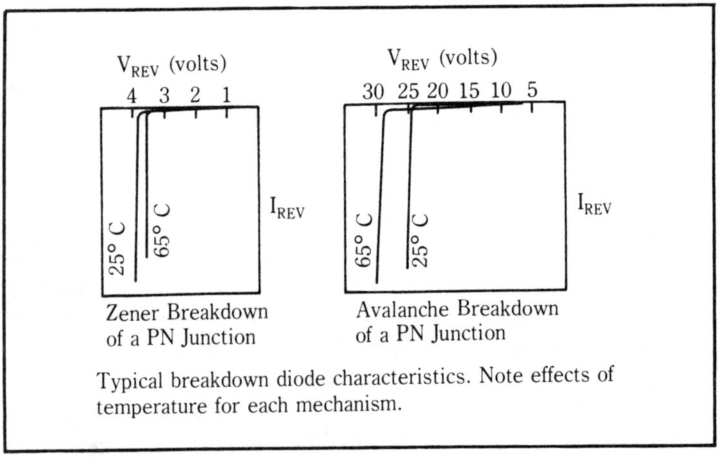

Fig. 2-3. Temperature effects on zener and avalanche diodes. (From *Motorola Zener Diode Handbook,* copyright 1967 by Motorola; reprinted with permission.)

volts, the avalanche characteristic predominates. Just as resistors have their resistance values specified with a typical 5 percent or 10 percent tolerance, zener diodes also have their zener breakdown voltages similarly specified. The very popular 1N746-1N759 series of zener diodes have suffixes denoting this. An "A" suffix means a 5 percent tolerance; no suffix means a 10 percent tolerance.

Temperature-Compensated Zeners

Figure 2-4 shows a typical zener diode's dc V-I (volt-ampere) curve. This curve can be improved to resemble Fig. 2-5 with temperature-compensated zener diodes. Technically, if the diode has a breakdown voltage above 8 volts, it should be called an avalanche diode; however, convention is so strong that most people yield to the practice of still calling it a zener. Basically, a temperature-compensated zener is no more than an avalanche diode (positive temperature coefficient) in series with a silicon diode (negative temperature coefficient). The two cathodes are tied together (Fig. 2-6). The negative and positive temperature coefficients effectively cancel one another. Figure 2-7 is a graph of both alloy-diffused and diffused-junction zeners and shows how many silicon diodes must be used in series with the zener in order to yield a zero temperature coefficient. Note that at approximately 9.8 volts, the diffused-junction zener requires three PN junction diodes made from silicon to yield a zero TC (temperature coefficient). A zero TC zener can be immediately identified by measuring its

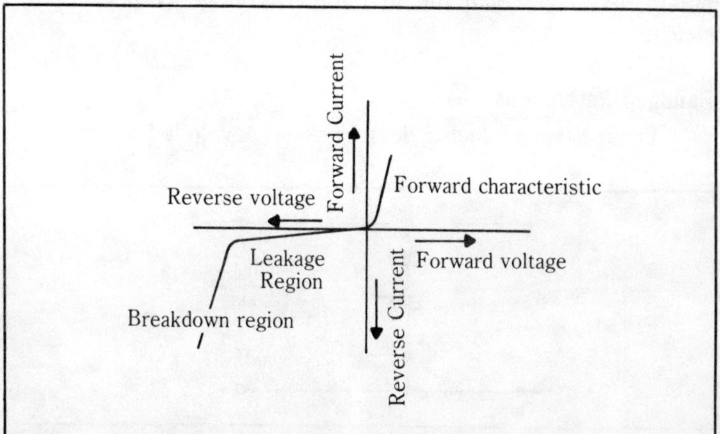

Fig. 2-4. Zener diode dc V-I characteristic curve. (from *Motorola Zener Diode Handbook*, copyright 1967, by Motorola; reprinted with permission.)

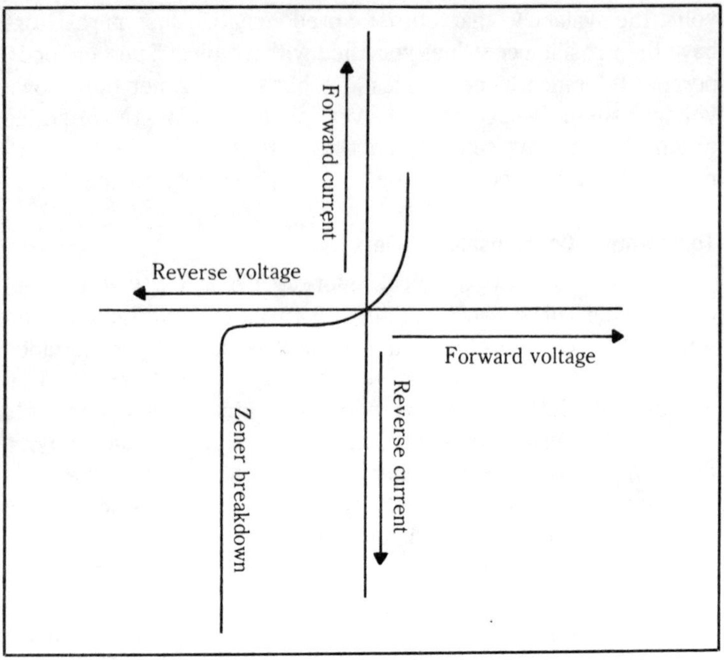

Fig. 2-5. Zener diode dc V-I characteristic curve (idealized).

resistance. The resistance should be quite high in both the forward and reverse direction.

However, a zener diode has two major drawbacks. It is very noisy electrically and the breakdown voltage (zener voltage) changes or *drifts*.

Bandgap References

These semiconductor devices were developed to overcome

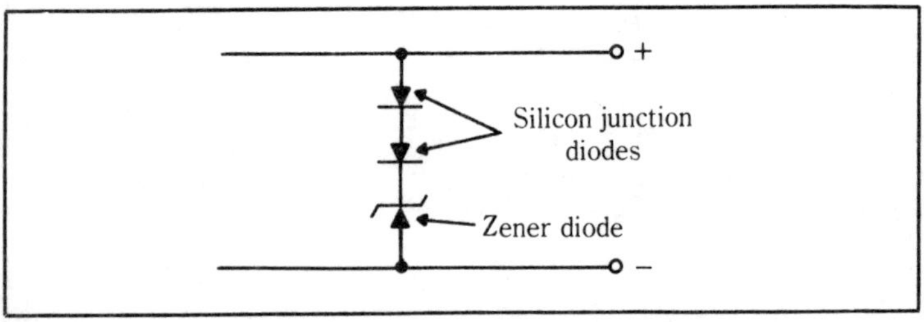

Fig. 2-6. Zener temperature compensation with silicon forward junctions. (From *Motorola Zener Diode Handbook*, copyright 1967 by Motorola; reprinted with permission.)

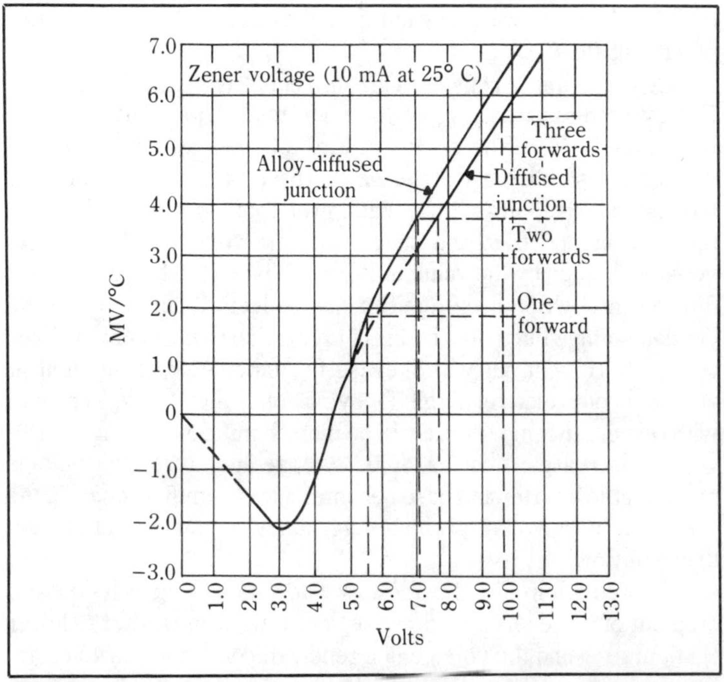

Fig. 2-7. Temperature compensation technique expressed in graphical form. (From *Motorola Zener Diode Handbook,* copyright 1967 by Motorola; reprinted with permission.)

the zener's poor noise qualities. Bandgap references are made from silicon and have theoretical potentials of 1.205 volts, which enables them to operate from supply voltages as low as 2.0 V. This is important for battery-powered and/or micropower devices. The other advantage of bandgap references is that they can be made to have very nearly zero temperature coefficient outputs. This type of voltage reference is now almost exclusively used in IC voltage regulator designs.

CONSTANT-CURRENT SOURCES

Constant-current sources are fundamentally important and are an integral part of a voltage regulator (Fig. 2-1). A constant-current source supplies an unvarying current to the voltage reference so that it will not vary due to changes in biasing current. A zener or bandgap voltage reference's performance is based upon receiving a constant, unvarying biasing current. These devices are schematically represented by either double intersecting circles or a single circle with an arrow coming out from the circle indicating direction of current flow.

A very simple constant-current source is shown in Fig. 2-8. However, this current source suffers from its dependence upon the regulation of V+. The circuit in Fig. 2-8 uses an R1/R2 voltage divider to establish a voltage on the base of the transistor. The transistor is used in an emitter follower configuration so that both the emitter and the base have the same voltage. This emitter voltage, divided by the value of R3, establishes the level of current flowing through both the emitter and collector. Since the R1/R2 divided voltage fluctuates in direct proportion to the supply voltage (V+), the current out will likewise fluctuate. Also, if the emitter voltage is not relatively high, (above 5 volts) then the V_{BE} changes with respect to temperature will be more significant. As the emitter voltage increases, this 0.1 or 0.2 volt temperature drift voltage between the emitter and base becomes a less significant part of the overall emitter voltage, which determines the amount of current that will flow.

The circuit in Fig. 2-9 allows V+ to vary as long as it does not drop out of its zener breakdown region. It uses an emitter follower configuration and the voltage of a zener/silicon diode pair to establish the base voltage and thereby, in turn, the emitter voltage which is divided by R3 to establish the magnitude of current. This circuit will be more resistant to voltage drift due to temperature changes since the zener and silicon diodes have opposite and cancelling temperature coefficients. The only requirement is that the voltage can't go below the zener voltage of the diode.

The two circuits in Figs. 2-10 and 2-11 use bandgap voltage

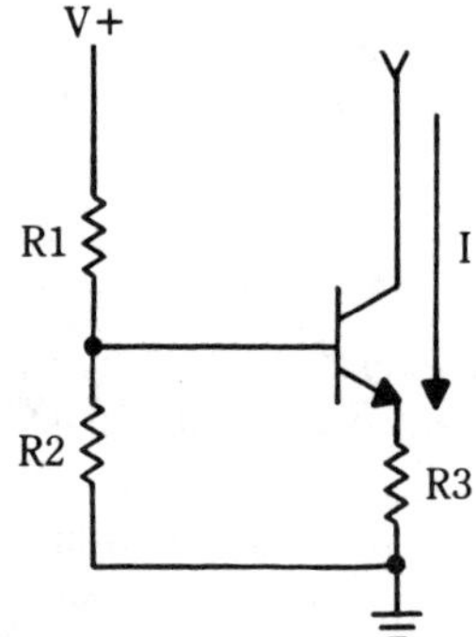

$$I = \frac{R2}{R1 + R2}\left(\frac{V+}{R3}\right)$$

Fig. 2-8. Simple constant-current source.

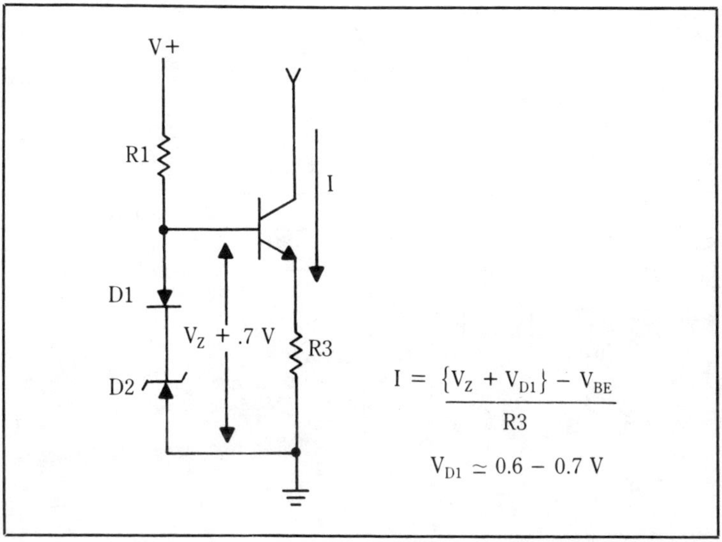

Fig. 2-9. Improved constant-current source with variable output.

reference constant-current sources. They supply a constant voltage across a fixed load so a constant current flows. The circuit in Fig. 2-10 consumes a scant 20 μA and operates from a 12 V source. The circuit uses an LM385 current source, a two terminal device, and a

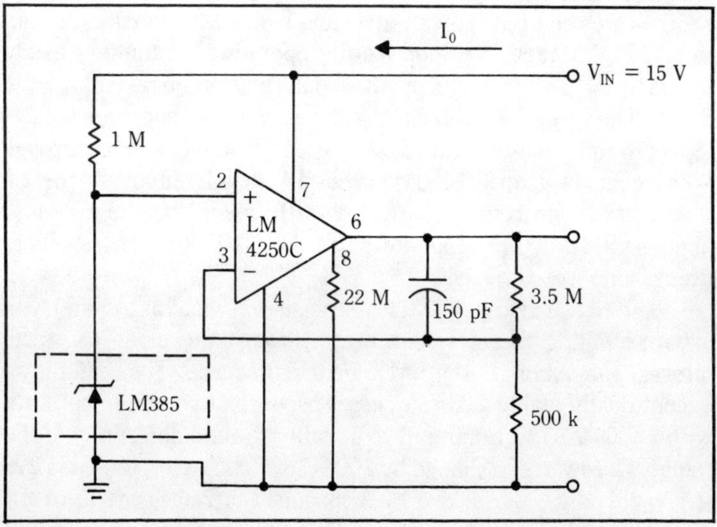

Fig. 2-10. Micropower precision 10.000 V reference. (From *EDN*, April 1, 1981, copyright 1981 by Cahner's Publishing, reprinted with permission.)

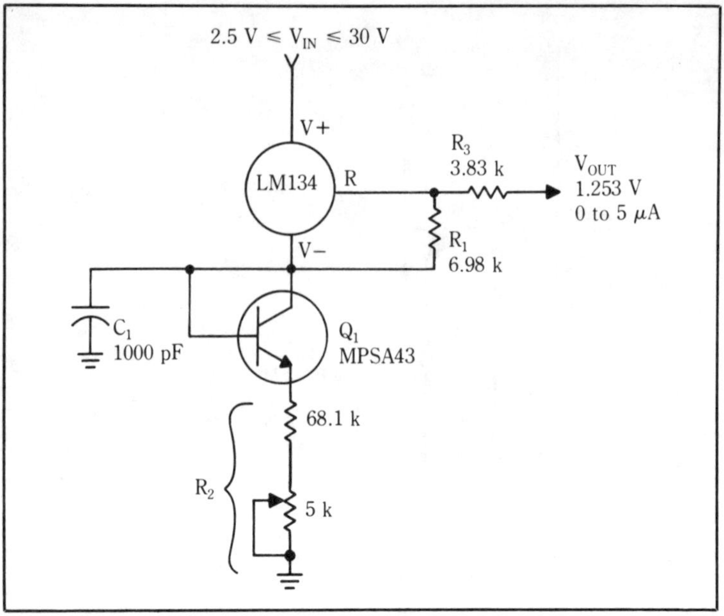

Fig. 2-11. Micropower precision trimmed reference. (From *EDN*, May 20, 1979, copyright 1979 by Cahner's Publishing; reprinted with permission.)

micropower programmable op amp, an LM4250C, to establish an extremely accurate voltage reference at 10.000 V. Such accurate sources are required in data converters with 10 or more bits resolution. The LM4250C op amp has its operating parameters established by the programming pin (8) and the size of the resistor placed there. The LM385 establishes a reference on the noninverting (+) input pin and is biased with 15 μA or the 15 V supply voltage through a 1 megohm resistor. The LM385 has a 1.25 volt output and the 1:8 feedback voltage from the 3.5 megohm/0.5 megohm voltage divider divides the 10.000 V output down to 1.25 V so that the two inputs to the op amp match.

The circuit in Fig. 2-11 uses a transistor (wired as a diode) with a pot trimming the current flowing through the diode to cancel internal gain errors in the LM134 current source. The result is an exceptionally stable 298 μA current source at room temperature with a 0.01 μA/°C temperature stability. Stated differently, if the temperature were to change by 100° C, which is a lot, then this 298 μA would only experience a 1 μA change in current—one-third of 1 percent—truly incredible! The LM134 has a positive temperature coefficient, and the silicon diode has a negative temperature coefficient. The amount of this negative temperature coefficient can be trimmed by the 5 k pot, establishing just the correct bias current so that the positive and negative temperature coefficients cancel one another. The results, as shown, are superb. The voltage at which this occurs is near the theoretical bandgap reference of silicon, 1.205 volts. In actual practice, the value usually falls between 1.20 and 1.28 volts. In this particular case, the actual value was 1.253 volts. Note in the previous circuit how it was assumed that the LM385 had a 1.25 volt output which is very nearly what we have here. About 5.5 percent of the total current flows out of the LM134 V– terminal; the remainder flows out of the R terminal. The amount of current out of the R terminal is determined by dividing 67.7 mV by the value of R1. The output voltage is 64 mV (the voltage between terminals R and V–) plus the diode voltage (the voltage from the collector/base to the emitter), plus the voltage across R2.

I have illustrated examples of bandgap references used to supply very accurate amounts of current. The actual bandgap principle is very complicated, but simply stated, a zero temperature coefficient is obtained by operating two very closely matched transistors at different collector current densities. The positive tem-

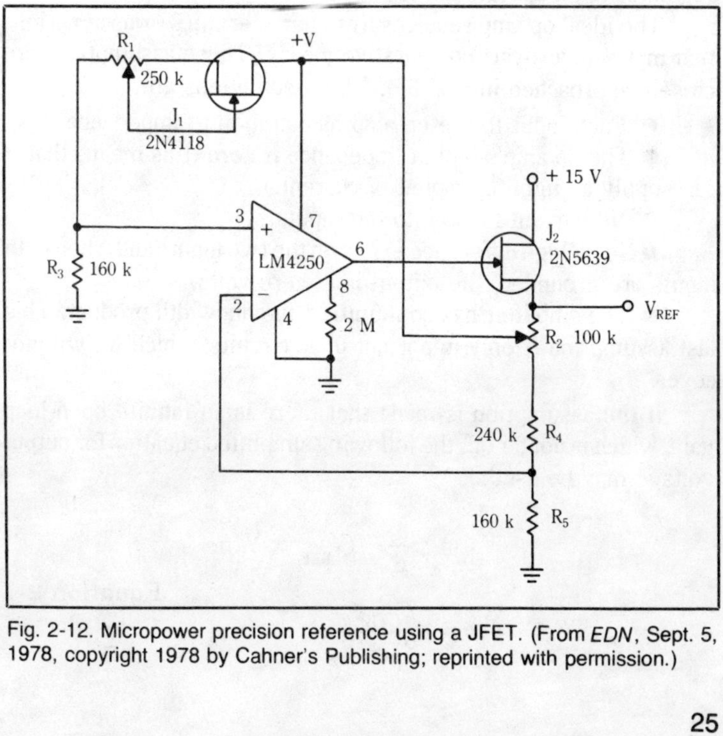

Fig. 2-12. Micropower precision reference using a JFET. (From *EDN*, Sept. 5, 1978, copyright 1978 by Cahner's Publishing; reprinted with permission.)

perature coefficient of the emitter-to-base voltage differential is added to the negative temperature coefficient of the base-to-emitter voltage of another transistor, and the two opposite coefficients cancel.

Figure 2-12 illustrates how a JFET and a micropower op amp can be used to make a low-drift, zero TC, voltage reference. A JFET can be biased slightly below its pinch-off voltage and at this point it exhibits a zero TC drain current (I_D). This circuit uses this principle to convert drain current to voltage by using an LM4250 micropower op amp. This op amp is current-programmable with respect to its operational parameters; however, JFET J1 allows the device's operation to be independent of the varying supply voltage and also current. The V_{REF} can be taken from the source of JFET J2 or from the wiper of pot R2. The purpose of pot R1 is to trim the very small TC resulting from the supply current of the op amp.

THE ERROR AMPLIFIER

Now that a stable reference has been established, the error amplifier must be discussed since it also effects the voltage regulator's performance. The error amplifier in a solid-state voltage regulator is an operational amplifier or "op amp".

The ideal op amp requires five characteristics or assumptions that make its analysis comparatively easy. These assumptions are closely approached in real life. The assumptions are:

- Each input to the op amp has an infinite impedance.
- The op amp's output impedance is zero (this means that it can supply an infinite amount of current.)
- No current flows into the inputs.
- No voltage difference exists on the two inputs and when both inputs are grounded, the output is at zero volts.
- The amplifier has an infinite gain-bandwidth product. This last assumption is only important in ac circuits, which we will not cover.

If the assumption is made that there is an infinite open-loop gain, which is not far off, the following simplified equation for output voltage may be used:

$$V_O = \frac{1}{\beta}(V_{REF} \pm V_{IO})$$

$$\beta = \frac{R_1}{R_1 + R_2}$$

Equation 2-1

where V_{REF} = reference voltage
V_{IO} = offset voltage
β = feedback or attenuation ratio
R1, R2 = network feedback resistors

The feedback or attenuation ratio is always less than 1, and IO equals the offset voltage. Thus, if the input offset voltage is small, the output voltage is merely a function of the voltage reference and the two resistors R1 and R2.

As shown in Equation 2-1, the term β is used with operational amplifiers to describe the voltage divider effect (attenuation) of the feedback resistor network of R1 and R2; however, when speaking of transistors, the term β is reserved to describe current gains of the transistor and the two should not be confused.

Error Amplifier Power Supply Sensitivity

Error amplifiers have both (+) positive and (−) negative supply voltage pins which are often denoted as V^+ and V^- respectively. Op amps are specified in different ways, however, many are specced as the total differential voltage across these pins. As an example, if an op amp were specced at +15 V and −15 V on the supply voltage pins, it would probably be safe to operate it at +25 V and −5 V, maintaining the 30 volt differential. Some op amps have their V^- pins grounded, so no ac output signal is possible with either voltage supply pin grounded. Check the data sheets before grounding either pin.

Changes in voltage regulator output voltage due to changes in power supply variations are caused by two main sources. The first is the power supply rejection ratio (PSRR). This is also called ripple rejection. With the use of a truly constant-current source, as is present in modern integrated voltage regulators, the PSRR problem can be ignored since it is so small. The second is caused by the common mode rejection ratio (CMRR). Common mode rejection is the ability of the op amp to reject signals that are present simultaneously at both inputs. The common mode rejection ratio, however, is a problem.

In an ideal op amp, only the differential voltage on the input pins should have an effect on the output; however, the common mode voltage also affects the output voltage. There are two methods of improving the op amp error amplifier by minimizing the CMRR. One is to make the open-loop gain as high as possible, but there are limitations to how high this gain can be made. One other

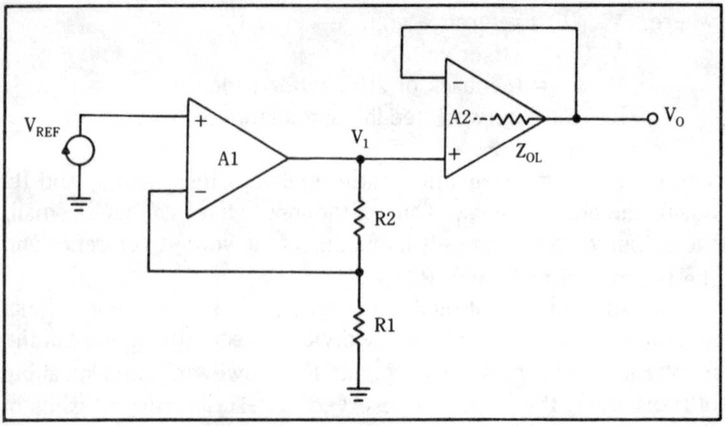

Fig. 2-13. The "regulator within a regulator" configuration. (From *Motorola Voltage Regulator Handbook*, copyright 1976 by Motorola; reprinted with permission.)

method for improving voltage regulator error amplifier and CMRR performance is for β to approach 1 or unity. However by making $\beta = 1$, there is another obvious problem, which is—how the designer can ever get the output voltage greater than V_{REF} with no error amplifier gain (which there obviously cannot be with $\beta = 1$). The answer lies in the regulator within a regulator approach. Refer to Fig. 2-13 and note how amplifier A1 establishes a reference voltage for amplifier A2 and how amplifier A2 has 100 percent feedback. This type of 100 percent feedback emitter-follower configuration is also called a unity-gain buffer and lessens the output impedance of amplifier A2 as well as the whole voltage regulator. Also, any voltage desirable may now be derived by the proper selection of the R1/R2 ratio.

THE SERIES-PASS TRANSISTOR

The next of the five elements constituting the modern voltage regulator is the internal series-pass transistor. This transistor, which is built on the actual integrated circuit, is usually in the emitter-follower configuration and is capable of passing somewhere between 20 mA and 1.5 A, although there are some recently introduced regulators that will be examined later that can carry 10 A. Let's assume that your circuitry needs more current than the voltage regulator can supply. This is a common problem and can be remedied by use of external discrete series-pass transistors.

In the following chapters, you will see a number of transistors used external to the IC voltage regulator. These transistors are either for current boosting and/or current limiting. They will be configured in the vast majority of cases as emitter followers. This means that the emitter "follows" the base or, more specifically, the emitter's voltage will match the voltage on the base of the transistor as long as the collector is more positive than the base. In addition to current boosting, the emitter follower also transforms impedances. That is, if you have one stage of amplification (a transistor) with an output impedance that is relatively high, then the stage you are driving must have an extremely high impedance (resistance) or the output voltage will be divided between two stages without the final driven stage seeing the full voltage developed by the driving or first stage. The emitter follower has an output with a very low impedance so that none or very little of the output voltage will be shared, but rather the vast majority of the developed voltage will appear at the input to the second stage. This voltage-sharing phenomenon we are trying to avoid is called "loading down a circuit" and the emitter follower prevents this from happening. In our applications, the first stage or transistor is the output transistor on the voltage regulator IC itself. This output transistor is always configured as an emitter follower with its collector tied to a positive source with no collector resistor used, its base driven internally, and its emitter with a resistor or another transistor as the load to ground. This is done purposely to avoid loading down a circuit. The second stage (transistor) has its base driven by this ICs emitter follower and often this second stage is even configured as an emitter follower itself.

Additionally, the beta of the series-pass transistor along with its maximum collector current are the two most crucial parameters with which we will be dealing. The beta of a transistor is the ratio of its collector current divided by its base current. As an example, in Chapter 5 a 1.8 A power supply will be designed using a 723 voltage regulator, which has a 150 mA maximum current capacity. Naturally, to attain 1.8 A of output current, an external series-pass transistor will be required. If 120 mA of current is supplied to the base of the external series-pass transistor by the 723 voltage regulator and the collector had a 1.8 A capacity, this is 15 times 120 mA so an external transistor with a minimum beta of 15 is required. In IC voltage regulator designs, you must first determine what the IC voltage regulator will handle in current and then divide your current requirement by this amount to get a number which represents the minimum beta required for the external series-pass transistor. Actually, to be safe, you should not push either the IC or the series-pass transistor to their current carrying limits because this continuously elevated level of operation severely degrades reliability and in turn lessens the life of the power supply. So, before you begin to design this "beefed up" version, be certain that the diodes in the rectifier bridge and that the transformer secondary can withstand these new current demands. Refer to the design problem which concludes Chapter 1.

The crucial parameters here are the current and power ratings of the PNP series-pass transistor. The $V_{IN} - V_{OUT}$ difference voltage across the transistor has to be multiplied by the current (maximum) that will flow through the transistor, and this product, expressed in watts, is the amount of power the transistor will have to dissipate. The collector current of the transistor should also be considered when looking at the data sheet.

In Fig. 2-14, a voltage regulator with a sense lead is shown. Some three-terminal regulators do not have sense leads and this will be covered later. In Fig. 2-14, the voltage regulator supplies the base current to the PNP series-pass transistor which acts as a current amplifier. A transistor in an emitter-follower configuration has its emitter follow the voltage on the base, or in this case, the output voltage of the voltage regulator. Figure 2-15 shows a PNP series-pass transistor used with a three-terminal voltage regulator. In Fig. 2-15, R, the base-to-emitter resistor, serves to route vol-

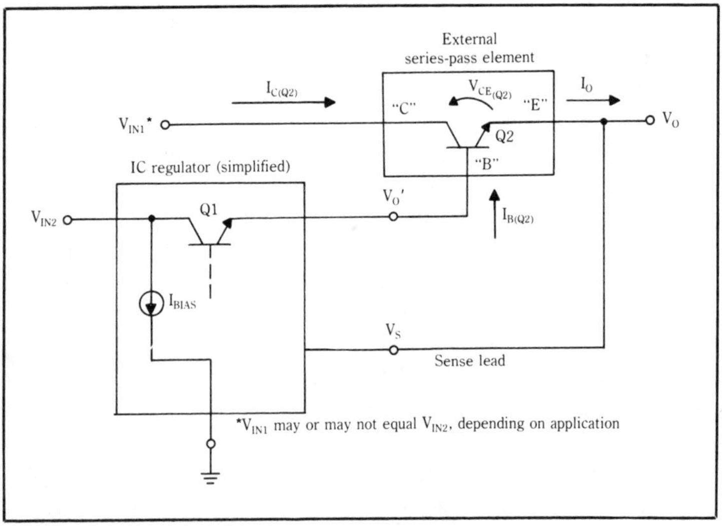

Fig. 2-14. NPN type series pass element configuration. (From *Motorola Voltage Regulator Handbook*, copyright 1976 by Motorola; reprinted with permission.)

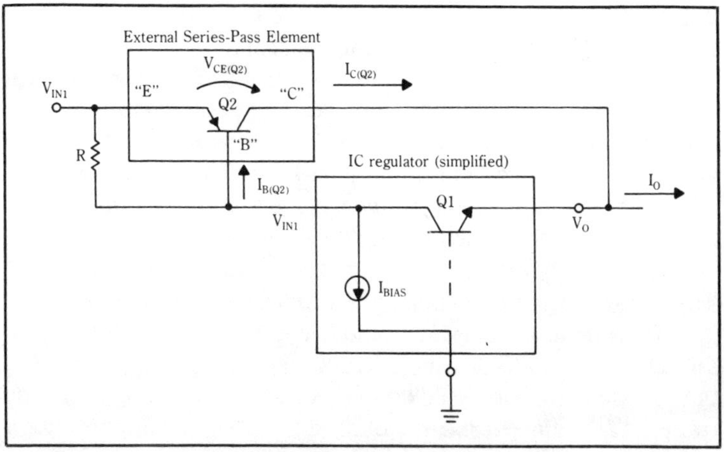

Fig. 2-15. PNP type series pass element configuration. (From *Motorola Voltage Regulator Handbook*, copyright 1976 by Motorola; reprinted with permission.)

regulator bias current away from the base of the series-pass transistor. This resistor should be low enough to prevent Q2 from turning on when I_{BIAS} flows. This relationship can be expressed by the following approximation:

$$0 < R_{BE} \leqslant \frac{V_{BE}\ \text{ON (Q2)}}{I_{BIAS}} \qquad \textbf{Equation 2-2}$$

where R_{BE} = base-to-emitter resistor
V_{BE} = base-to-emitter voltage
I_{BIAS} = bias current

THE INTERNAL VOLTAGE REGULATOR PROTECTION CIRCUITRY

The purpose of this circuitry is to prevent the voltage regulator from being damaged by operation at elevated temperatures in either very hot ambient environments or from excessive heat caused by high currents drawn from the voltage regulator's internal output drive transistor. The internal series-pass transistor can be operated at elevated currents above the recommended maximums if the integrated circuit's substrate is kept cool. How to avoid this self-heating phenomenon is covered in Chapter 8, the thermodynamics chapter.

Refer to Fig. 2-16, showing the basic voltage regulator with protection circuitry and note the thermal shutdown transistor. The voltage from the base-to-emitter of a transistor decreases as heat increases. The thermal shutdown transistor, Q3, is also located physically on the IC next to the series-pass transistor (the heat producer) so it directly senses heat. The base of Q3 is held at approximately 0.4 V which is below the transistor's turn on voltage at room temperature. As the integrated circuit's substrate or internal circuitry begins to heat up, the voltage to turn on Q3 decreases, eventually reaching a point when 0.4 V turns it on. When it does, this removes the base drive from Q1, the series-pass transistor. The voltage regulator is then turned off, and it cools until the 0.4 V is sufficiently low again and can no longer turn on Q3. However, if this turn-on/turn-off threshold was in the neighborhood of 0.1° F, the voltage regulator would oscillate on and off, creating switching transients throughout the system and resulting in an intermittent square wave voltage. This situation has been avoided by internal circuitry on the voltage regulator chip known as *thermal hysteresis* circuitry. This allows a much wider range of turn-on/turn-off temperatures to occur. As an example, National regulators typically turn off at between 150° C and 190° C and do not turn back on for another 10° C less than that.

EXPERIMENT 2-1

This experiment demonstrates how to obtain a voltage source

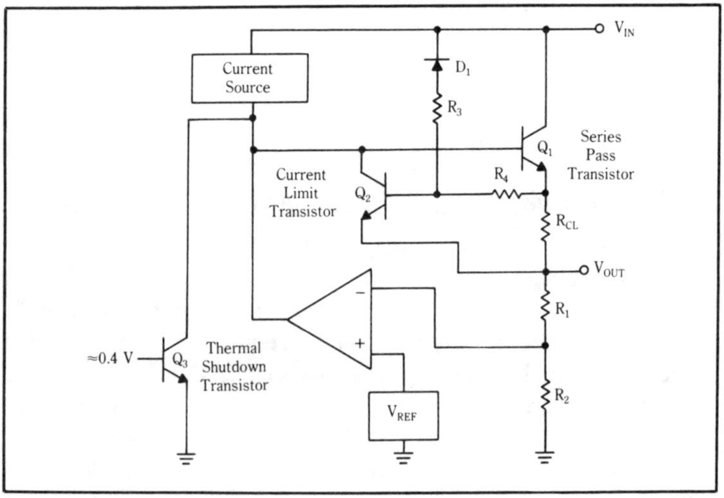

Fig. 2-16. Basic regulator with protection circuit. (From *National Semiconductor Voltage Regulator Handbook*, copyright 1980 by National Semiconductor; reprinted with permission.)

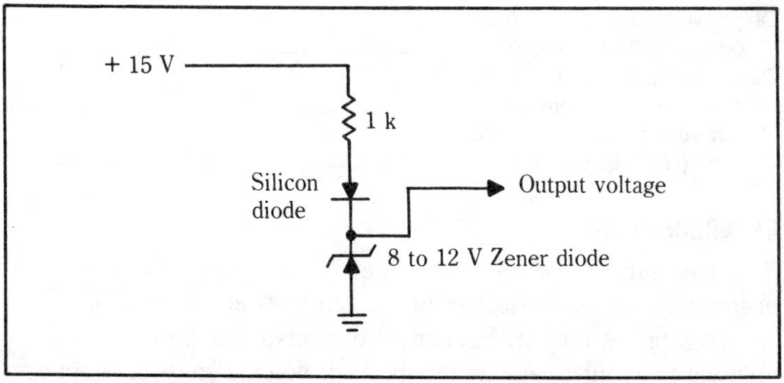

Fig. 2-17. Adding silicon diode in series with a zener diode to obtain zero temperature coefficient voltage source.

with a zero temperature coefficient by adding silicon diodes in series with a zener diode.

1. First wire the circuit as shown in Fig. 2-17 with only one diode in series with the zener diode, attach a DVM (10 Vdc scale) and record the voltage at point A on the data sheet at room temperature.

2. Turn on a heat gun or blow dryer and apply heat toward the junction of the zener and silicon diode. Record the new value under the HOT-1 diode column on the data sheet.

3. Repeat step 2 with two silicon diodes in series with the zener. Record the value of output voltage under the HOT-2 diodes column on the data sheet.

4. Repeat step 2 with three silicon diodes and record under the appropriate column on the data sheet. You may wish to add additional silicon diodes. The result is that as additional diodes are added, the temperature coefficient approaches zero or actually starts to become negative. This is where you can "trim" the number for maximum effectiveness.

Data Sheet

Output voltage with one silicon diode (room temperature) ________________

Output voltage with one silicon diode (HOT-1-diode) ________________

Output voltage with two silicon diodes (room temperature) ________________

Output voltage with two silicon diodes (HOT-2-diodes) ________________

Output voltage with three silicon diodes (room temperature) ________________

Output voltage with three silicon diodes (HOT-3-diodes) ____________________

Output voltage with four silicon diodes (room temperature) ____________________

Output voltage with four silicon diodes (HOT-4-diodes) ____________________

EXPERIMENT 2-2

The purpose of this experiment is threefold. First, it demonstrates how an emitter follower can help eliminate ripple in a power supply's output. Second, it demonstrates how the emitter follower is a control element with the dc voltage on its base appearing at the emitter. Last, it shows how the source supplying the drive to the base need not have anywhere near the current drive capability as that of the output stage provided by the transistor's beta which current amplifies the base-to-collector currents by an amount equal to the transistor's beta.

1. Connect the circuit shown in Fig. 2-18.
2. Set the vertical amplifier of an oscilloscope to 0.5 V/cm or

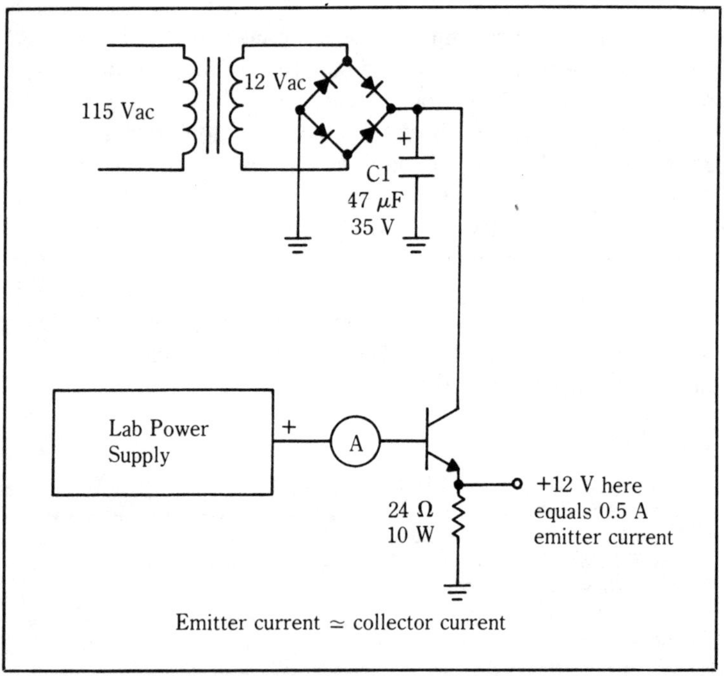

Fig. 2-18. Multipurpose emitter-follower serves as ripple-filter, current amplifier and dc voltage control.

50 mV/cm if you are using a 10:1 probe. Set the time to 10 ms/cm division.

3. Measure the ripple on the collector of the power transistor.

4. Measure the ripple on the emitter of the power transistor.

5. Take the ratio of these two figures and note that as long as the collector voltage does not fall below the base voltage that the emitter is almost impervious to ripple. Also, if desired, this allows the regulator, which consumes much less power, to be supplied by a regulated low-power source and allows less filtering and a less well-regulated voltage source to supply the power series-pass transistor.

6. Vary the supply to the base and note how the emitter voltage "follows" this base voltage.

7. Increase the base voltage to a point that is above the lower negative-going peaks of the series-pass transistor supply voltage's ripple and note the emitter's waveform. It should be a continuous positive dc voltage with small portions of the waveform dipping slightly at a rate of 60 Hz.

8. With an ammeter in series with the base current, determine how much base current needs to flow for the series-pass transistor's emitter to put out 0.5 A. The ratio of these currents is the beta of the series-pass transistor.

Chapter 3

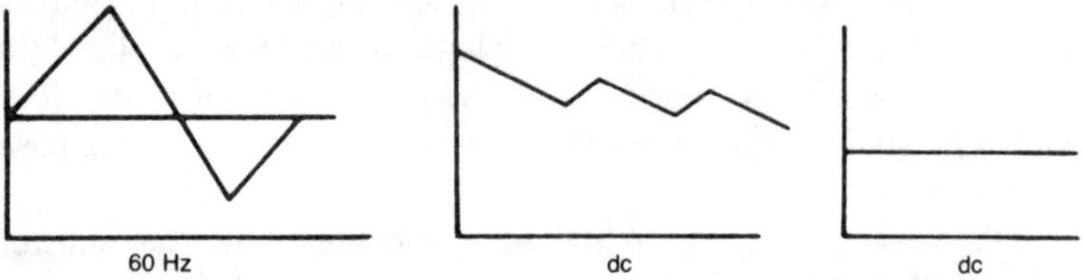

Types of Linear Voltage Regulators

Voltage regulators fall into four main categories (Table 3-1). First, there are the three-terminal fixed-voltage types with the 7800 series positive and 7900 series negative regulators used here as representative examples. Second, there are three- and four-terminal adjustable voltage regulators spanning a 2 V to 37 V range. The LM117 is used in this chapter as the primary example within this category. Third, there are switching voltage regulators. Chapter 4 covers this subject with three representative IC switching voltage regulators. Last, there are precision multiterminal regulators, of which the LM723, LM100, LM104, and LM105 are typical examples. The LM723 alone is the subject of Chapter 5. There is a subcategory within this type of regulator. This is the multiple-output dual-tracking type like the LM125 and LM126 regulators. These two are identical in function, but the first has a ± 15 V output while the second has a ± 12 V output. The LM125 is covered in this chapter with the understanding that the circuits shown using it are equally applicable to the LM126.

FIXED VOLTAGE REGULATORS

A fixed-voltage regulator is one with a constant or nonvarying output voltage that is determined at the time of manufacture. Typical output voltages include both negative and positive values of 5, 6, 8, 10, 12, 15, 18, and 24 volts.

The most common type of fixed-output voltage regulator is the three-terminal type. Table 3-1 shows these regulators with output voltage and current ratings. Table 3-2 shows a comparison of

Table 3-1. Regulator Output Current and Voltage Listing. From *National Semiconductor Voltage Regulator Handbook*, copyright 1982 by National Semiconductor; reprinted with permission.)

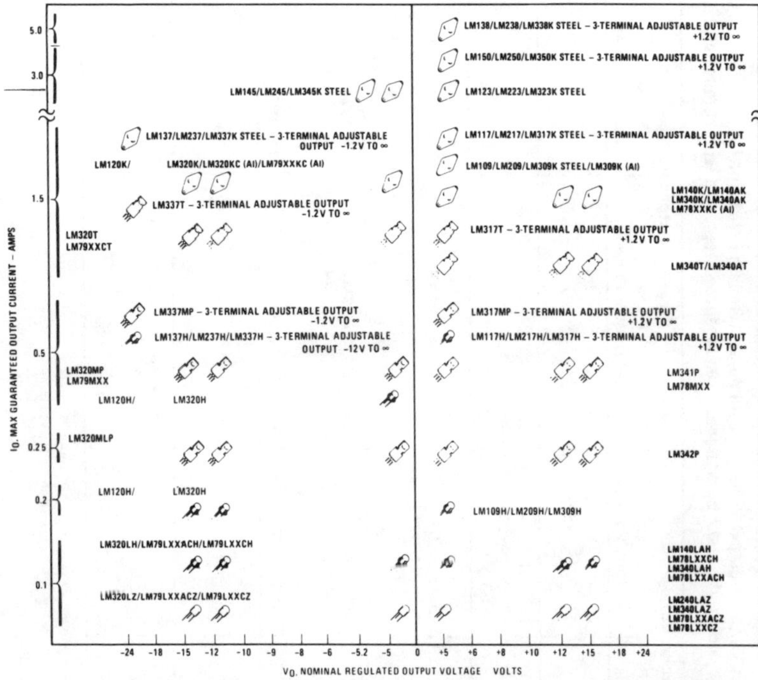

Note: All devices with TO-3 package designation (K and K STEEL) are supplied in steel TO-3 packages unless otherwise designated as (Al) aluminum TO-3 package. All devices with KC package designation are supplied in aluminum TO-3.

fixed-voltage three-terminal regulators and variable-voltage regulators by use and features.

Figure 3-1 shows a typical 7800 series three-terminal voltage regulator with a bypass capacitor and single-point grounding. Figure 3-2 shows how an output current in excess of the regulator's maximum rating can be obtained. Resistor R1 determines the threshold point at which the transistor begins to conduct and is given by the relationship within the illustration. The beta of the transistor must be greater than ten for this relationship to remain valid. The transistor in Fig. 3-2 is not short-circuit protected; however, this can be accomplished by the circuit shown in Fig. 3-3. This circuit has a short-circuit resistor R_{SC} in the series bypass path to sense current. When transistor Q1 is bypassed, the regulator goes into the short-circuit mode. Most voltage regulators by themselves have a short-circuit protection circuit built right on the chip but when operated at current levels greater than the specified

Table 3-2. Fixed Three-Terminal vs Variable-Output Voltage Regulators. (From the *National Semiconductor Voltage Regulator Handbook*, copyright 1982 by National Semiconductor; reprinted with permission.)

Output Current		Positive Output Voltage: Fixed Output Voltage	Positive Output Voltage: Adjustable[2] Output Voltage	Negative Output Voltage: Fixed Output Voltage	Negative Output Voltage: Adjustable[2] Output Voltage
5 Amp	Device		LM338		
	Output Voltage		+1.2V to +33V		
	Package		TO-3		
3 Amp	Device	LM323	LM350	LM345	
	Output Voltage	+5.0V	+1.2V to +33V	−5.0V, −5.2V	
	Package	TO-3	TO-3	TO-3	
1.5 Amp	Device	LM340-XX, LM78XX	LM317	LM320-XX, LM79XX	LM337
	Output Voltage	+5V, +12V, +15V	+1.2V to +37V High Voltage (HV) +1.2V to +57V	−5.0V, −12V, −15V	−1.2V to −37V High Voltage (HV) −1.2V to −47V
	Package	TO-3, TO-220	TO-3, TO-220	TO-3, TO-220	TO-3, TO-220
0.5 Amp	Device	LM341-XX, LM78MXX	LM317M	LM320M, LM79MXX	LM337M
	Output Voltage	+5V, +12V, +15V	+1.2V to +37V	−5.0V, −12V, −15V	−1.2V to −37V
	Package	TO-202	TO-202, TO-39	TO-202, TO-39[1]	TO-202, TO-39
0.25 Amp	Device	LM342-XX		LM320ML	
	Output Voltage	+5V, +12V, +15V		−5.0V, −12V, −15V	
	Package	TO-202		TO-202	
0.10 Amp	Device	LM340LA-XX, LM78L-XX		LM320L-XX, LM79L-XX	
	Output Voltage	+5V, +12V, +15V		−5V, −12V, −15V	
	Package	TO-39, TO-92		TO-92, TO-39	

Note 1: Some voltage options are rated only to 200 mA.

Note 2: Adjustable voltage regulators can regulate voltages to infinity.

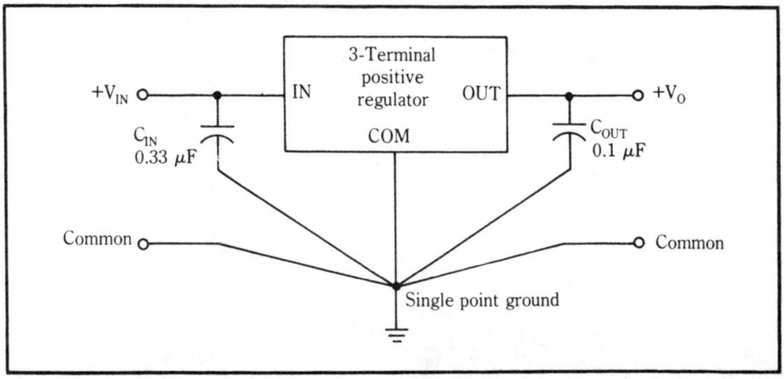

Fig. 3-1. Basic three-terminal regulator with bypass capacitors (From the *Fairchild Voltage Regulator Handbook,* copyright 1978 by Fairchild; reprinted with permission.)

maximum, a separate external short-circuit protection circuit must be designed, as is done here.

Another method of short-circuit protection is shown in Fig. 3-4. The two resistors, R1 and R2, set the ratio of current handled by the series-pass transistor and voltage regulator. This circuit can have its current-carrying capability extended to that of the series-pass transistor itself by proper selection of regulator and transistor heatsinks. This will be shown in Chapter 8. The circuit in Fig. 3-4 does have the drawback that the voltage drop across the current-sharing resistors, R1 and R2, causes the circuit dropout voltage to be considerably higher. The circuit in Fig. 3-5 shows how this problem is solved with both a negative and a positive three-terminal voltage regulator.

Voltage regulators often experience fluctuations of as great as ±30 percent in the current limiting capability of the IC itself. The

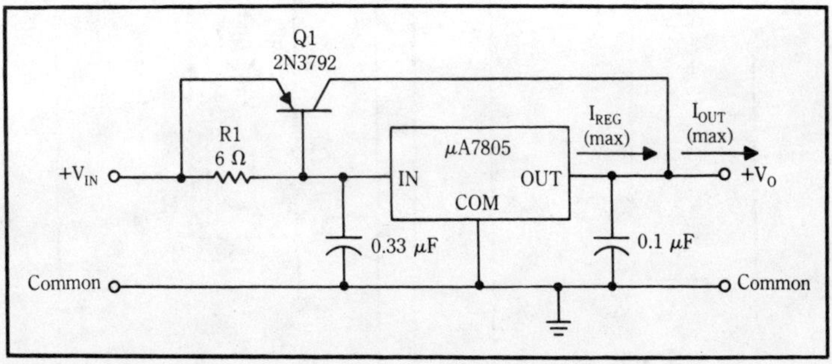

Fig. 3-2. High-current voltage regulator. (From the *Fairchild Voltage Regulator Handbook*, copyright 1978 by Fairchild; reprinted with permission.)

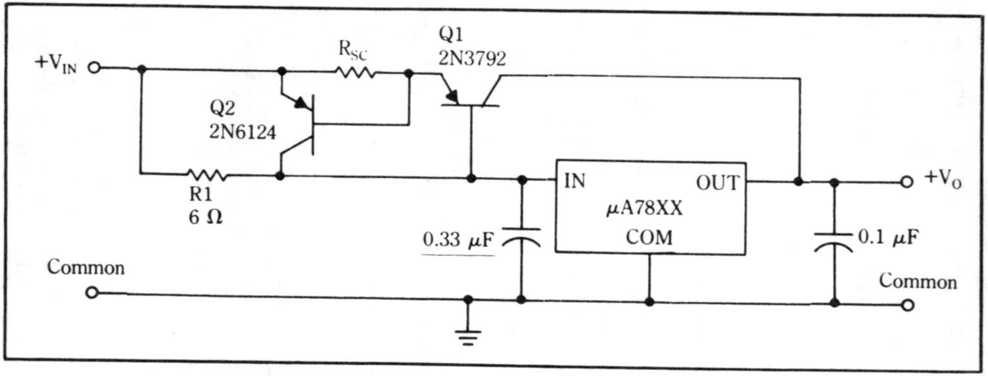

Fig. 3-3. High-current voltage regulator with short-circuit protection. (From the *Fairchild Voltage Regulator Handbook*, copyright 1978 by Fairchild; reprinted with permission.)

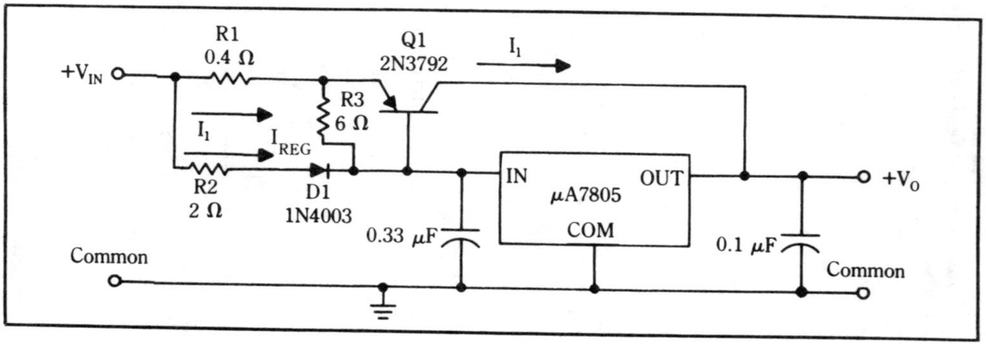

Fig. 3-4. Five-amp regulator with short-circuit protection. (From the *Fairchild Voltage Regulator Handbook*, copyright 1978 by Fairchild; reprinted with permission.)

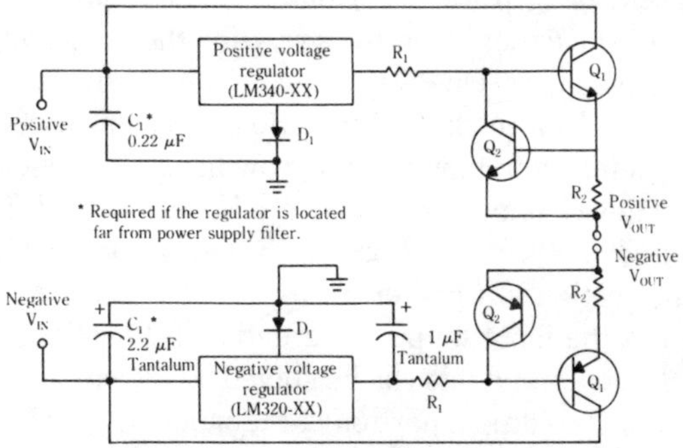

$$I_{SC} = \frac{V_R + V_{D1} - V_{be2} - V_{be1}}{R_1} + \frac{V_{be2}}{R_2}$$

where $\frac{I_{LOAD\ MAX}}{\beta_{1\ MIN}} < I_{SC}$ for the regulator

V_R = fixed regulator voltage

I_{SC} = short-circuit current

Fig. 3-5. Discrete current limiting with three-terminal regulators. (Copyright 1978 by *EDN*; reprinted with permission.)

circuit in Fig. 3-5 allows for rather precise limits to be set at which the current is limited. This value may be set above or even below the rated maximum current limit of the regulator itself. This circuit reduces the problem of heatsinking because in these particular regulators a maximum of 125° C to 150° C is allowed. The maximum temperature allowed in the series-pass transistors is 200° C. Both current limiters become active when their load currents become high enough to cause the voltage drop across R2 to equal the base-to-emitter voltage of Q2. The purpose of Q2 is to limit the base drive on Q1 so that its output current is also limited. The diode in the ground leg of these regulators compensates for the loss in output voltage (V_{BE}) and also for temperature drift from Q1's V_{BE}.

If the use of series-pass external transistors is not desired, a scheme using regulators in parallel can be used to enhance the overall current capability of the circuit. This is shown in Fig. 3-6. Regulators operating in parallel, however, require their output voltages to be very nearly equal (better than two percent matching is required). If one regulator has a greater output voltage, it supplies the majority of current to the load and this can be catastrophic! The unequal current division from multiple regulators in parallel results in the major portion of current being supplied by only *one* regulator. This poses the possibility of the particular "overworked" regulator having its current capability exceeded. If it fails, the remaining parallel regulators must supply the full load current, a load current often too great for the remaining reduced number of regulators. One way to avoid this is to match regulator output voltages by a quick hand selection or grading process using a "window" detector to monitor a narrow band of acceptable output voltages. These voltages are visually indicated by LEDs (light-emitting diodes) (refer to Figs. 6-18 and 6-19 and the window comparator section in Chapter 6).

At times the input-output differential of the voltage regulator has to be exceeded and this can be solved by the circuit in Fig. 3-7. Here R1 serves to drop a portion of the input voltage to lower the input-output differential voltage across the regulator to a point that is within the limits of the regulator. This selection of R1 is important so that the regulator does not drop out of regulation under worse case conditions of maximum output voltage and load and minimum supply voltage. The equation for R1 is in the illustration. Another slightly more complex circuit to accomplish this is shown in Fig.

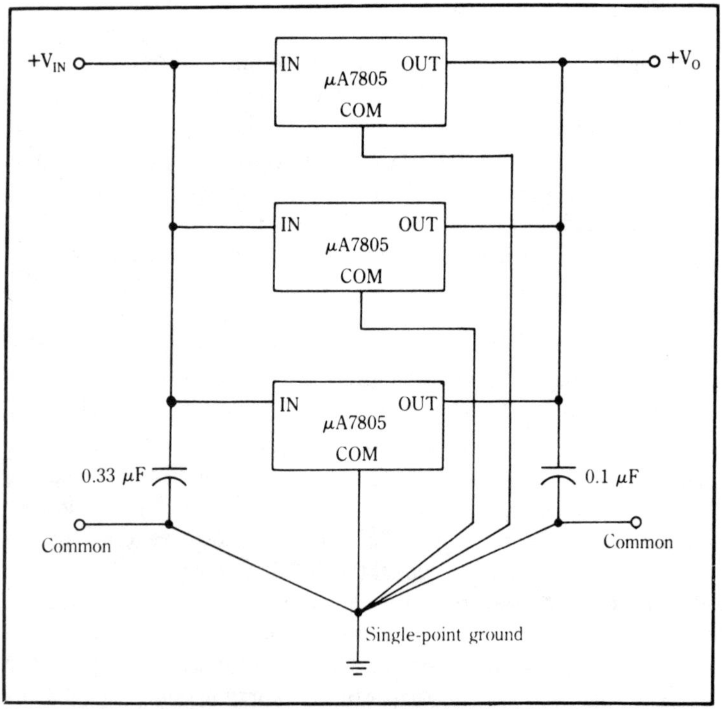

Fig. 3-6. Paralleling regulators for current boosting. (From the *Fairchild Voltage Regulator Handbook*, copyright 1978 by Fairchild; reprinted with permission.)

$V_{S(MIN)}$ = the minimum supply voltage
$V_{DO(MAX)}$ = the maximum dropout voltage $\simeq$ 3.0 V
$I_{Q(MAX)}$ = the maximum quiescent current $\simeq$ 5 mA

$$R1 \simeq \frac{V_{S(MIN)} - V_{O(MAX)} - V_{DO(MAX)}}{I_{OUT(MAX)} - I_{Q(MAX)}}$$

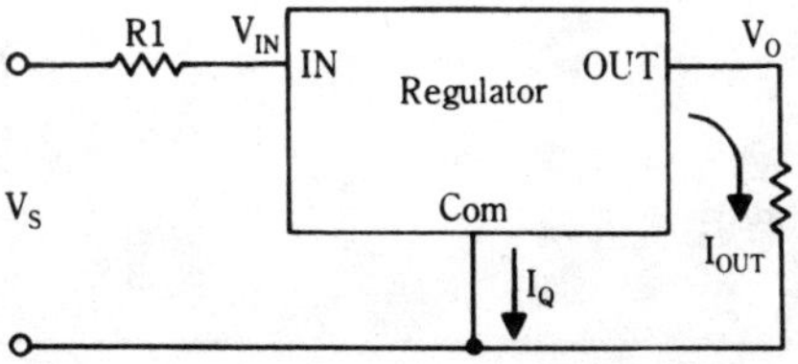

Fig. 3-7. Reducing regulator power dissipation. (From the *Fairchild Voltage Regulator Handbook,* copyright 1978 by Fairchild; reprinted with permission.)

3-8. Here the regulator is supplied with a constant voltage while the input voltage may vary.

It is often desirable to generate an output voltage greater than that of the fixed voltage of the regulator. The circuit in Fig. 3-9 shows a simple method of "tricking" the voltage regulator by using a zener diode to move the voltage up by the value of the zener's breakdown voltage. More specifically, the regulator in this instance is a +24 V model; however, the desired output voltage is +44 V, so obviously we must place the regulator's common or ground pin at a potential of +20 V above ground. This is precisely what has been done here. There is a drawback to this "quick and dirty" design procedure. If the input voltage exceeds the maximum input voltage rating of the regulator, the thermal and short-circuit protective

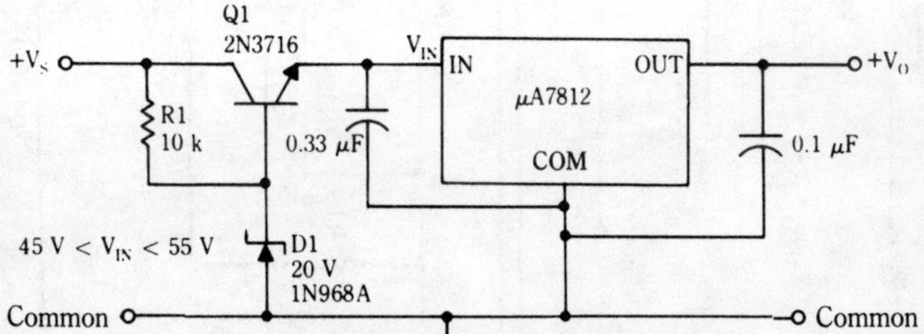

Fig. 3-8. High-input voltage regulator with protection. (From the *Fairchild Voltage Regulator Handbook,* copyright 1978 by Fairchild; reprinted with permission.)

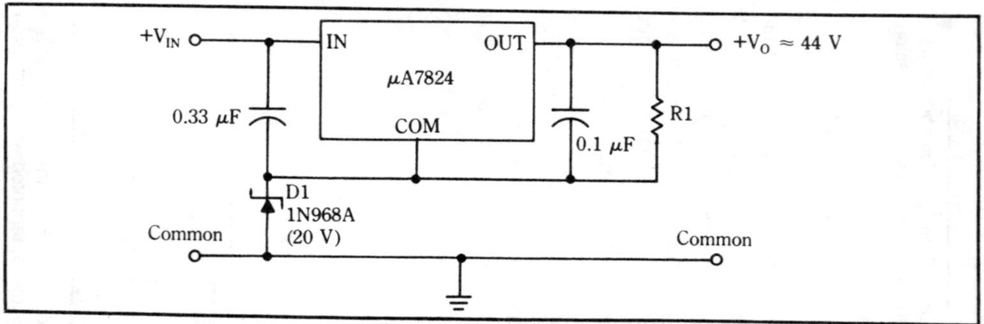

Fig. 3-9. High output voltage regulator. (From the *Fairchild Voltage Regulator Handbook*, copyright 1978 by Fairchild; reprinted with permission.)

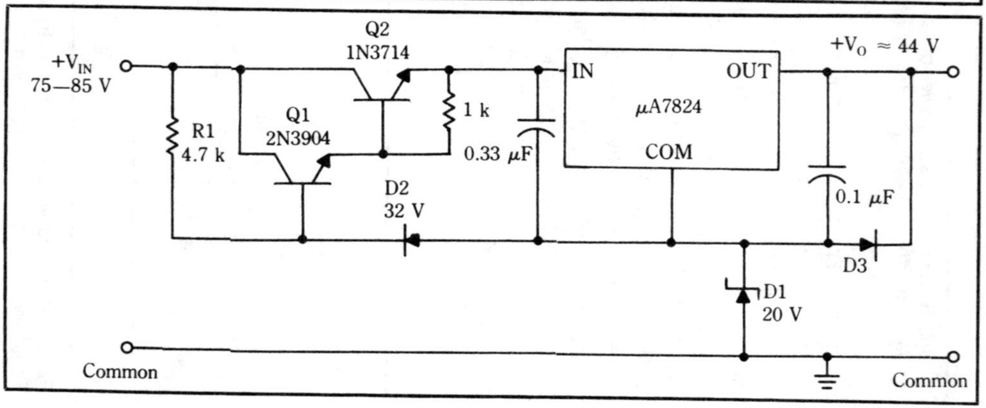

Fig. 3-10. High output voltage regulator with short-circuit protection. (From the *Fairchild Voltage Regulator Handbook*, copyright 1978 by Fairchild; reprinted with permission.)

features built into the regulator are nullified. To solve this problem and still maintain these protective features while achieving elevated levels of output voltage greater than that specified in the voltage regulator data sheets, the circuit in Fig. 3-10 can be used. This circuit uses a Darlington pair (transistors Q1 and Q2) to reduce the input voltage to a safe level. It also prevents excessive loading on zener diode D2.

Occasionally you may want to shut down the regulator yourself rather than relying on the current-limiting short-circuit protection circuitry or the internal thermal protection circuitry of the regulator. This electronic shutdown may be initiated either manually or by a logic level driving a transistor. A typical application of electronic shutdown would be when you wanted to remove power from the load. Electronic shutdown of regulators like the 723, 104, and 105 is relatively easy because the noninverting input of the error amplifier is accessible or brought out to a pin on the IC package. Four-terminal regulators have their control terminal as the inverting input to the error amplifier so a few added external components are necessary to accomplish electronic shutdown. The three-terminal regulators have the same arrangement as the four-terminal regulators so an arrangement as shown in Fig. 3-11 is used. Here, Q2 is normally on and provides base current to Q1. When Q2 is turned off, the base to Q1 no longer receives any base current and shuts off the regulator. The turn-off time of the regulator is a function of C2 and the load current with faster turn-off times occurring with increased load currents.

As stated previously, a three-terminal regulator has a fixed output voltage set at the factory; however, it is possible to get a voltage other than one of these nonstandard voltages by use of the

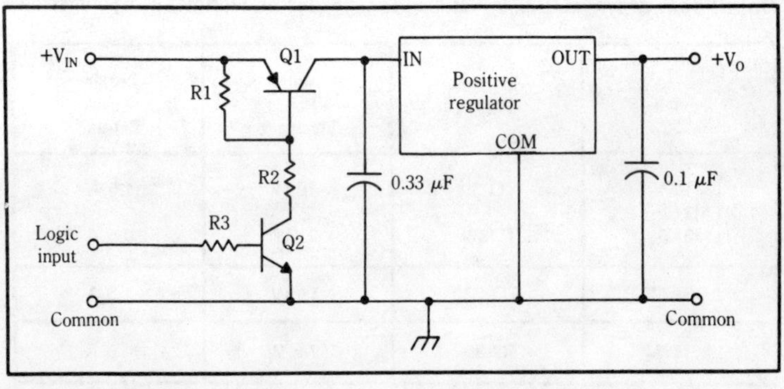

Fig. 3-11. Remote or electronic shutdown. (From the *Fairchild Voltage Regulator Handbook*, copyright 1978 by Fairchild; reprinted with permission.)

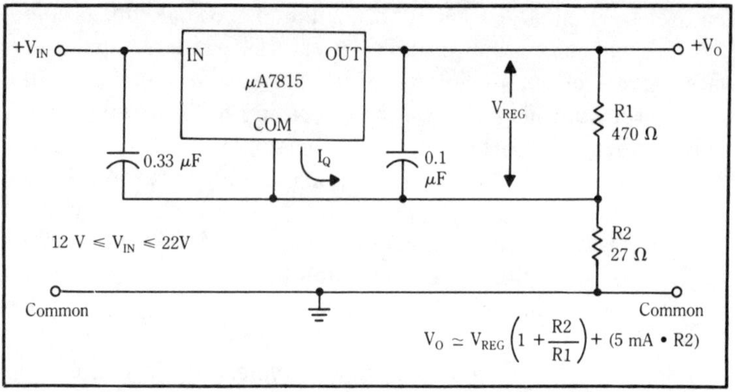

Fig. 3-12. Nonstandard fixed-output voltage regulator. (From the *Fairchild Voltage Regulator Handbook,* coypright 1978 by Fairchild; reprinted with permission.)

circuit in Fig. 3-12. Note that the output voltage is determined, among other factors, by the quiescent current of the regulator itself, which varies with temperature. However, this term's contribution to the overall output voltage fortunately is rather slight.

ADJUSTABLE THREE-TERMINAL REGULATORS

The LM117 and the LM137 negative complement regulator to the LM117 are two of the most popular adjustable voltage regulators. The adjustable or "programmable" output voltage allows its use in applications formerly limited only to discrete designs.

Table 3-3 shows the LM117 series of packages and power

Table 3-3. LM117 Package and Power Options. (From the *1982 National Semiconductor Voltage Regulator Handbook,* copyright 1982 by National Semiconductor; reprinted with permission.)

Device	Package	Rated Power Dissipation	Design Load Current
LM117 LM217 LM317	TO-3 TO-39	20 W 2 W	1.5 A 0.5 A
LM317T	TO-220	15 W	1.5 A
LM317M	TO-202	7.5 W	0.5 A
LM317LZ	TO-92	0.6 W	0.1 A

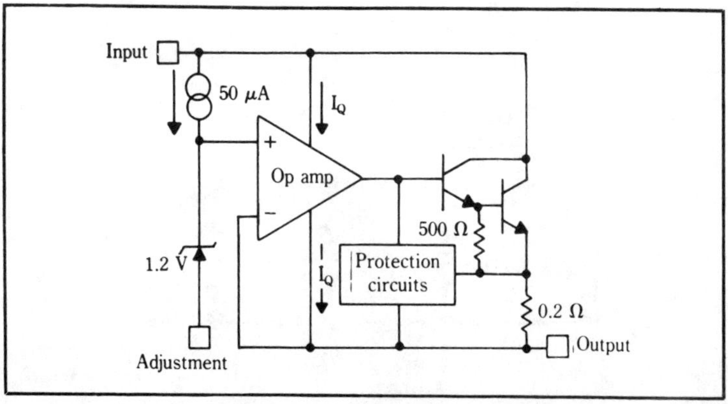

Fig. 3-13. LM117 block diagram. (Courtesy National Semiconductor.)

options available. Incidentally, National Semiconductor uses a numbering convention of prefixing its parts with either a 1, 2, or 3 to indicate military, industrial, and commercial grade components, respectively. Therefore, an LM117 and LM137 are identical with the exception that the 117 has tighter specs and an extended operating temperature range. The LM117 family yields 0.01 percent/V linc regulation and 0.1 percent load regulation. The LM117 does not need to be referenced to ground.

A block diagram of the voltage regulator is shown in Fig. 3-13. A data sheet on the LM117 is found in the Appendix. The output voltage is accurate to within a ±3 percent tolerance and can be trimmed with the arrangement shown in Fig. 3-14. Another arrangement that accomplishes trimming but does so with discrete

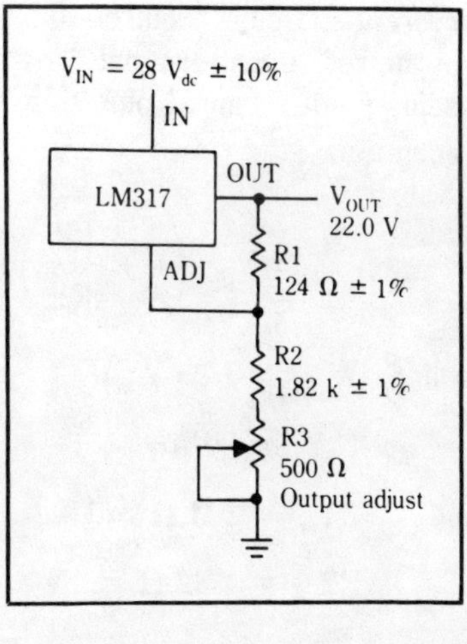

Fig. 3-14. Trimming the LM117 with a pot. (Courtesy National Semiconductor.)

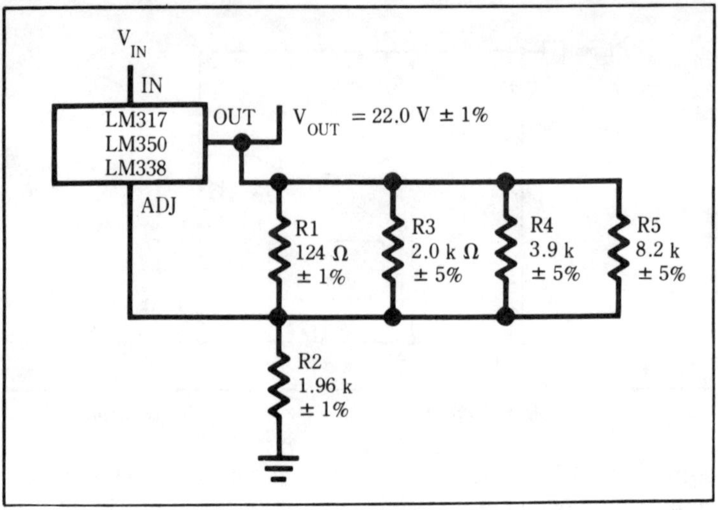

Fig. 3-15. Trimming the LM117 with discrete resistors. (Courtesy National Semiconductor.)

components to avoid a costly pot is shown in Fig. 3-15. Note how R3, R4, and R5 are just common 5 percent tolerance resistors and how their values double each time as they go to the right. This adding of larger and larger value paralleled resistors each time causes an even smaller resulting overall resistance which is the same as "fine tuning" a pot by decreasing resistance each time. Observe in Fig. 3-13 how all quiescent current is delivered to the output rather than ground so that a separate ground terminal is not needed. About 50 μA of bias current from the regulator's constant current source is required to bias the 1.25 V reference and appears at the reference terminal. The output voltage is the voltage of the adjustment terminal plus this 1.25 V described by the following equation:

$$V_{OUT} = V_{REF}\left(1 + \frac{R2}{R1}\right) + I_{ADJ} \times R2$$ **Equation 3-1**

where $V_{REF} = 1.25$ V

$I_{ADJ} = 50\ \mu A$

$R1 = 124\ \Omega \pm 1\%$

$R2 = 1.82\ \Omega \pm 1\%$

The floating nature of this regulator requires all the quiescent current to be absorbed by the load; therefore, too light a load impairs good regulation. A "bleeder" resistor drawing 10 mA should therefore always be present on the output. The LM117 current-limiting circuitry is internally set at 2.2 A and remains constant with temperature. Older regulators like the 309 and 7800 series regulators used the turn-on of an emitter-base junction to set the current limit. This junction voltage is temperature sensitive as previously pointed out and causes the current limit to change over a 2:1 ratio from −55° C to +150° C. This is avoided with the LM117.

This regulator can have improved rippled rejection with the insertion of two capacitors as shown in Fig. 3-16; also note the diodes. These are for protection because when the capacitors discharge, they may find a path of discharge through low-current junctions in the IC and may damage it. When shorted, the average 10 μF capacitor has an internal resistance low enough to allow 20 A spikes to be generated.

The circuit in Fig. 3-17 is required for a true zero output since the adjustment terminal must be 1.2 V below ground to offset the internal voltage reference.

The LM117, 217, or 317 programmable voltage regulators make ideal candidates for high-voltage power supplies because of their floating-mode operation; however, the $V_{IN} - V_{OUT}$ differential limitation of 40 volts presents a problem. The circuit in Fig. 3-18

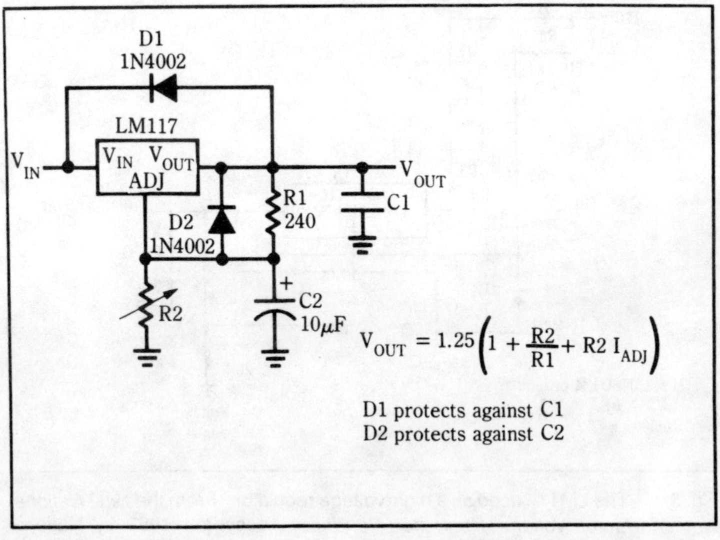

Fig. 3-16. Inserting diode protection. (Courtesy National Semiconductor.)

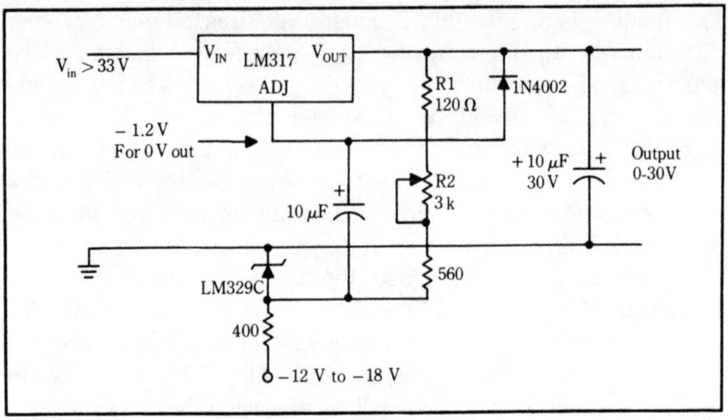

Fig. 3-17. Obtaining a true zero output with the LM117.

solves this problem by allowing the Q1, Q2 Darlington pair of power transistors to absorb this differential. These low-beta transistors allow for dead shorts across the regulator over the entire range of output voltages from 12 V to 160 V. The LM329B precision zener diode is in series with the regulator's internal reference to enhance temperature stability and regulation. This is a temperature-compensated zener.

The LM117, 217, and 317 have a shutdown feature. The reg-

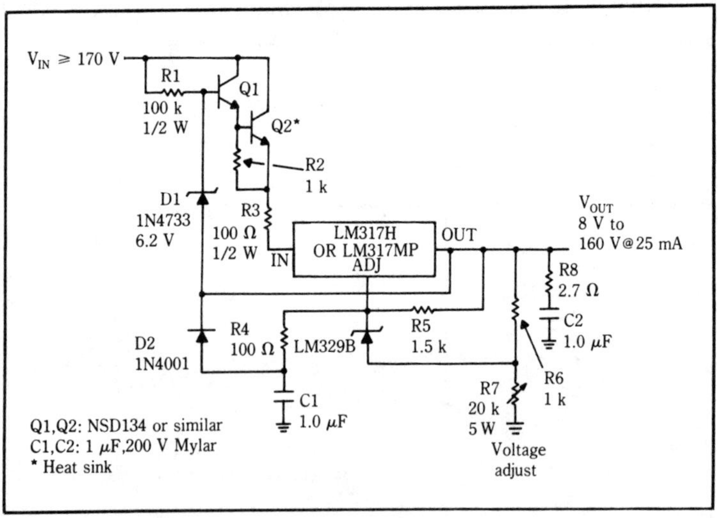

Fig. 3-18. The LM117 used as a high-voltage regulator. (From the *1982 National Semiconductor Voltage Regulator Handbook*, copyright 1982 by National Semiconductor; reprinted with permission.)

ulator shuts down to a 1.25 V output, the reference voltage, when the adjustment pin is shorted to ground through a transistor. The circuit in Fig. 3-19 uses this ability plus the built-in clamping and voltage-sensing ability of a normally on JFET for a simple power-on sequencing circuit for dynamic RAMs. Dynamic RAMs, like the popular 16K × 1 bit 4116 RAM, require that the V_{BB} (−5 V) not be present when the V_{DD} (+12 V) is first applied or that V_{BB} never be more positive than V_{SS} (ground) when V_{DD} is first applied. Also, V_{BB} must never be more than 0.2 V positive with respect to V_{SS} when V_{DD} is applied. This circuit provides protection for these dynamic RAMs by sensing the absence of the −5 V supply and turning off the +12 V supply. When the JFET grounds the adjustment terminal, this 1.25 V on the V_{DD} of the dynamic RAMs powers them down and prevents damage by ensuring that the negative supply never floats slightly positive. An LM117 can deliver 1.5 A if packaged in a TO-3 IC. This is enough to easily power a 64K memory system.

Figure 3-20 shows an LM117 used with numerous common-emitter transistors whose bases are driven by logic levels. The inputs control how many transistors are turned on. The turning on of a transistor places its collector resistor in parallel with the resistor designated R2. Therefore, the more transistors turned on, the more resistors will be paralleled with R2, lowering the equivalent resistance value of R2 and resulting in a lower output voltage. There is a small drop from the collector to the emitter of a conducting transistor but this is only about 0.2 V and can be ignored in most calculations.

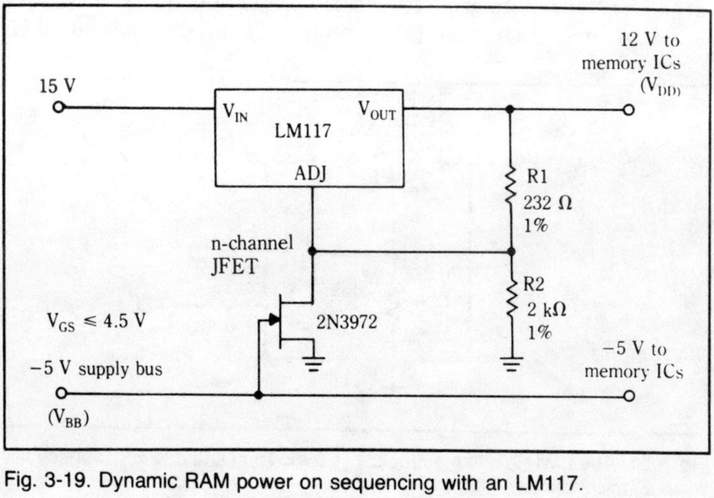

Fig. 3-19. Dynamic RAM power on sequencing with an LM117.

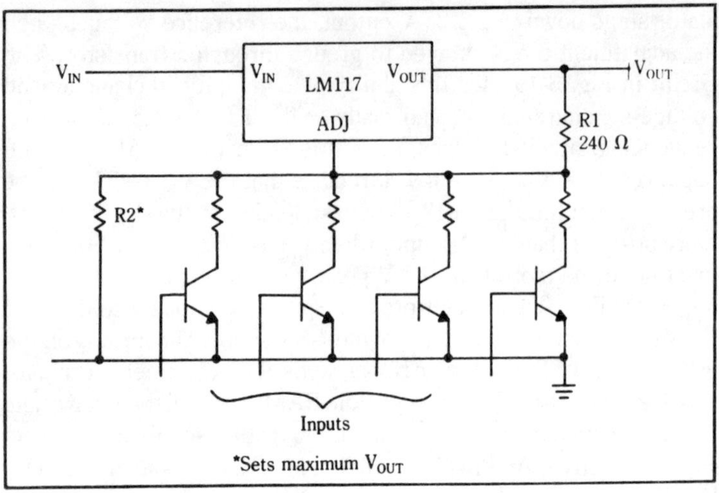

Fig. 3-20. The LM117 with digitally programmed output voltages. (Courtesy National Semiconductor.)

Figure 3-21 shows an LM117 with electronic shutdown actuated by a TTL logic level. When the TTL input is a logic 1, the transistor conducts and causes the LM117 adjust terminal to go low, diverting bias current through the transistor and causing the regulator to shut down. Figure 3-22 shows how to obtain multiple output voltages. The multiple output voltage circuit uses the constant programming current of the LM117 with the output setting resistor tapped to provide multiple output voltages to op-amp buffers. The transistors are for overload protection. When the output of the op amps is shorted, the op amps will current limit and their

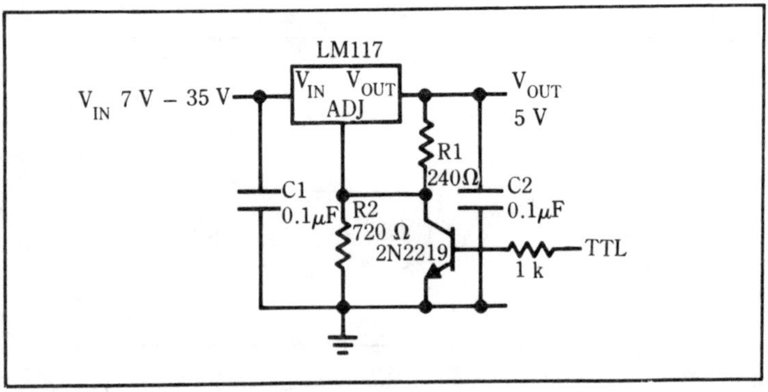

Fig. 3-21. The LM117 with electronic TTL logic level shutdown. (Courtesy National Semiconductor.)

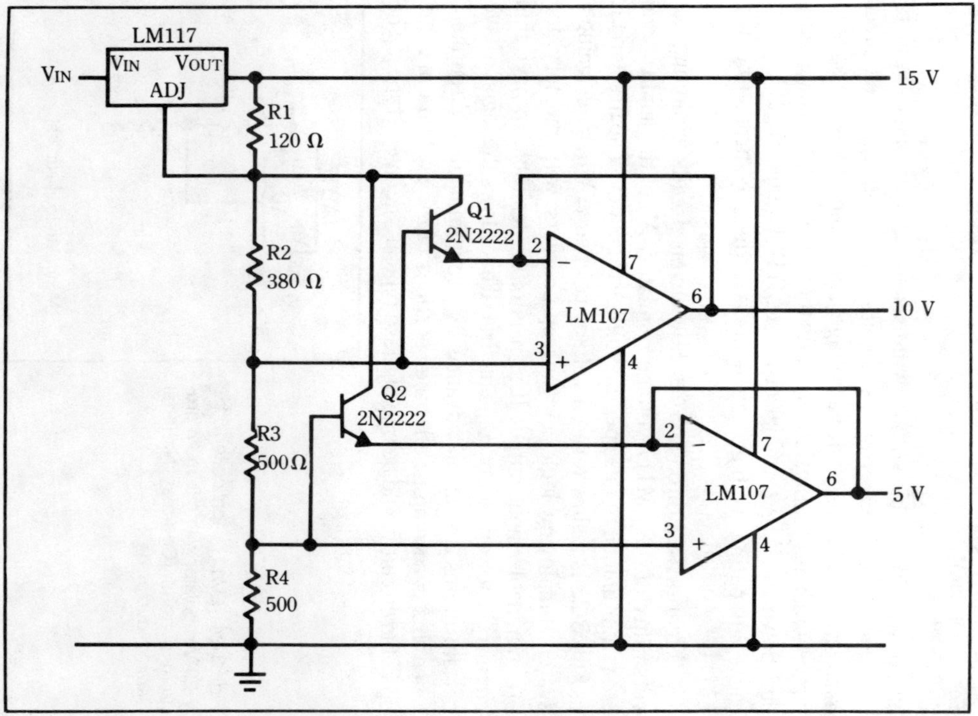

Fig. 3-22. Obtaining multiple-output voltages with the LM117. (Courtesy National Semiconductor.)

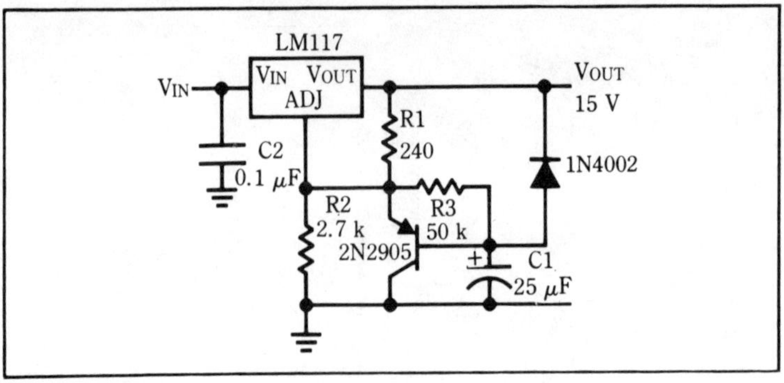

Fig. 3-23. LM117 application: slow turn-on 15 V regulator. (Courtesy National Semiconductor.)

output will drop. This voltage will make their inverting (−) inputs less than their noninverting (+) inputs. This is an undesired effect in an op amp and this positive voltage across the (+) to (−) op amp inputs will bias the transistors on. As they turn on and conduct, they pull down the ADJ terminal of the LM117 voltage regulator and cause all output voltages to decrease, thereby protecting the LM117.

Figures 3-23 through 3-29 show seven circuits exploiting the versatility of the LM117 voltage regulator. All of the applications are rather unique to this specific voltage regulator. The first circuit in Fig. 3-23 is a slow turn-on voltage regulator. When designing a system employing logic and microprocessors with memory and programmable input-output ICs, it is often desirable to have this circuitry's power supply come on first. This allows the logic to have certain states set up in advance by programmable subroutines. Once this is accomplished, other portions of the system can come up. This circuit would be one of these typical "delayed" types. The

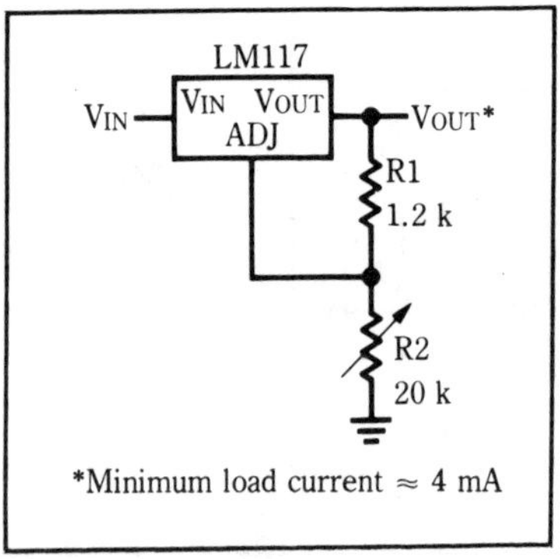

Fig. 3-24. LM117 application: 1.2 V-20 V regulator with minimum program current. (Courtesy National Semiconductor.)

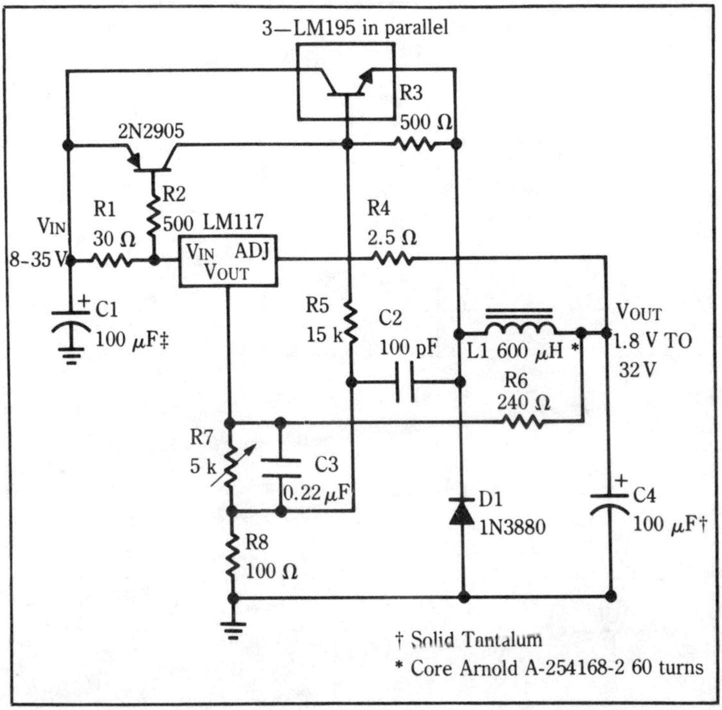

Fig. 3-25. LM117 application: 4A switching regulator with overload protection. (Courtesy National Semiconductor.)

+15 V regulator shown in Fig. 3-23 uses an emitter follower again, but it may appear "upside down" to you since it is a PNP transistor. When the LM117 is initially turned on, its output must first charge the 25 μF capacitor before the base of Q1 will have any voltage on it. Once this happens, the emitter will have an equal voltage on it (the output voltage). The time before full voltage is realized is very nearly five times constants or five times (25 microseconds) (50,000) = 12.5 milliseconds. If a longer time constant is desired, use a

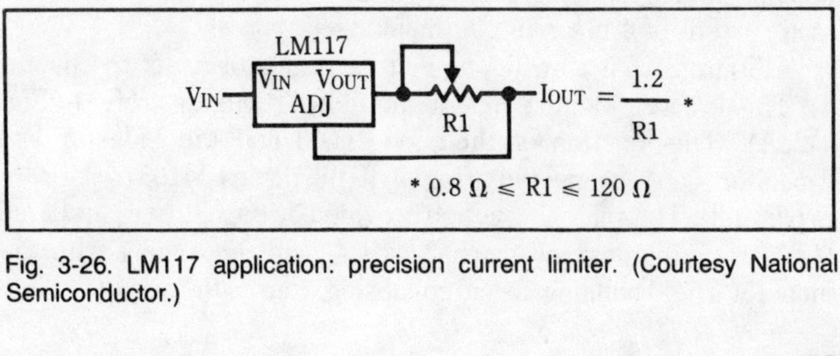

Fig. 3-26. LM117 application: precision current limiter. (Courtesy National Semiconductor.)

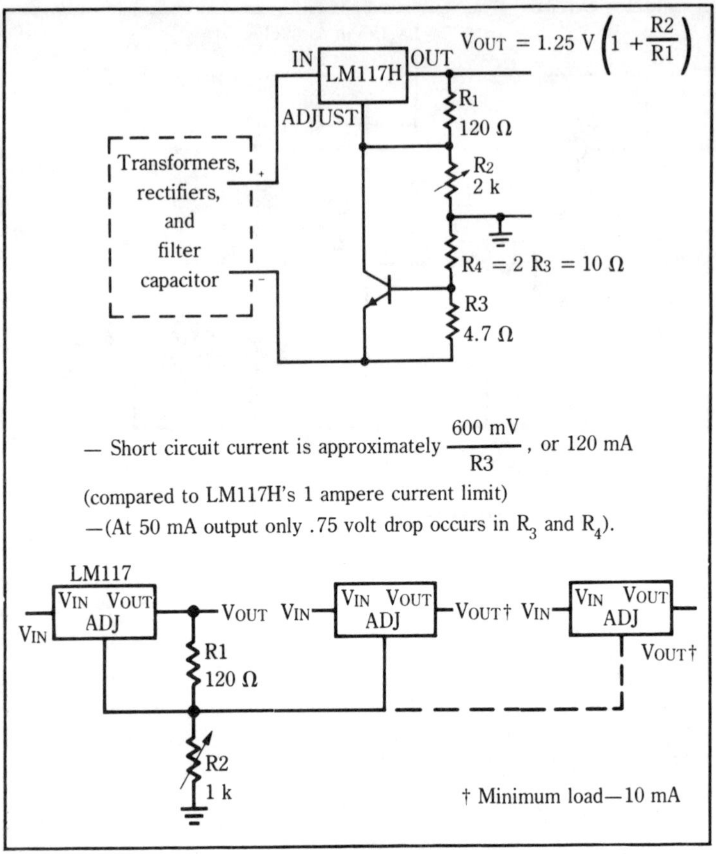

Fig. 3-27. LM117 application: adjusting multiple on-card regulators with single control (all outputs within ± 100 mV). (Courtesy National Semiconductor.)

larger capacitor. This is more expensive than using a larger resistor but a certain minimum no load current flowing out of the LM117 must always be maintained.

Figure 3-24 is a 1.2 V to 20 V regulator using a minimum amount of programming current. In this case again, the minimum load current of 4 mA must be maintained.

Figure 3-25 is a switching voltage regulator (refer to Chapter 4). This is a 4 A, highly efficient step-down regulator (refer to Fig. 4-25A). The junction of the diode (D1) and the inductor and capacitor (C2) is connected to a switch (the three LM195 transistors in parallel). The switching action provided by these three paralleled transistors charges and discharges C4, and this, along with the inductor's field building up and collapsing, causes the output voltage

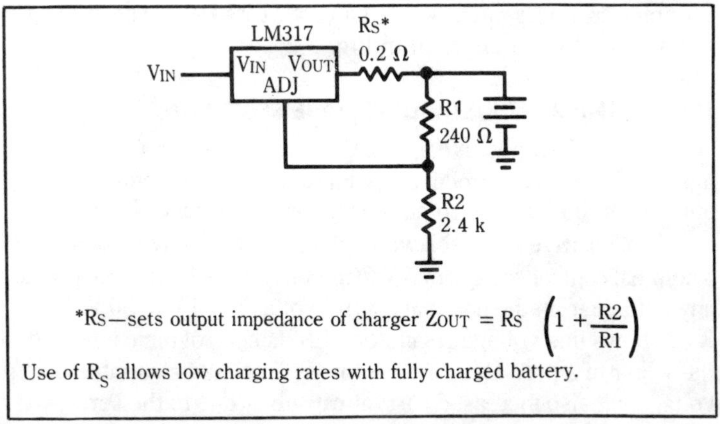

Fig. 3-28. LM117 application: 12 V battery charger. (Courtesy National Semiconductor.)

to be generated. Switching regulators are efficient, but their ripple is poor.

Figure 3-26 is a precision current limiter with the current limit equal to the internal voltage reference at approximately 1.2 V divided by the series output resistor, R1.

Figure 3-27 is a handy application for use with multiple on-card regulators. The circuit at the top uses an NPN transistor to establish a true ground. The multiple LM117 voltage regulators at the bottom are tied into this circuit with the intersection of R2 and R4 being the ground in the circuit to the right. The R1 and R2 resistors are the same in each circuit. They are only shown in more detail in the circuit to the left.

The circuits in Figs. 3-28 and 3-29 show two different versions of battery chargers. The 12 V charger uses the series limiting resistor R_s to establish an output impedance which in turn controls the charging rate. The lower the impedance, the faster the charging rate. The 50 mA battery charger provides a small current, commonly called a ''trickle" charge. This circuit is nothing more than a

Fig. 3-29. LM117 application: 50 mA constant-current battery charger. (Courtesy National Semiconductor.)

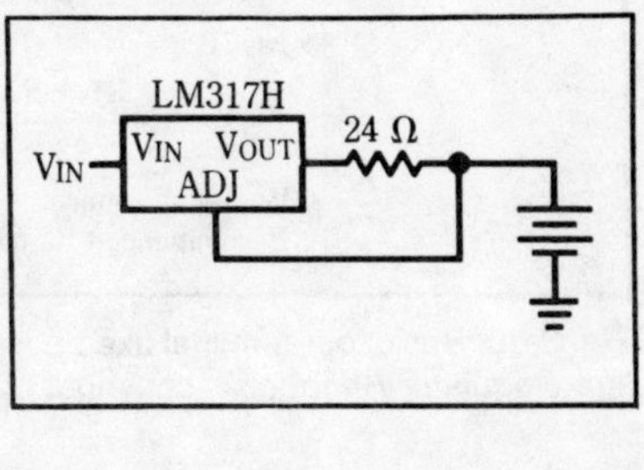

constant-current source with a battery as its load. The current is determined by the equation in Fig. 3-26.

FOUR-TERMINAL ADJUSTABLE VOLTAGE REGULATORS

The μA78G positive and μA79G negative four-terminal voltage regulator series produced by Fairchild has the traditional three pins (input, output, and common) but also has a control pin as well.

The positive adjustable voltage regulator, the A78G series, has a nominal control voltage of 5 V (Fig. 3-30). This circuit can provide any voltage from the nominal control voltage of 5 V to 30 V. Figure 3-31 shows this voltage regulator with its control pin connected to the wiper of a pot. This circuit provides not only a variable output voltage but also increased current capability due to the series-pass transistor, Q1. Transistor Q1 is in the feedback loop and therefore does not affect the temperature coefficient of the output voltage.

The negative four-terminal adjustable voltage regulator, the μA79G series, has a nominal −2.23 V reference. Refer to Fig. 3-32 for a typical application. Like its positive counterpart, the μA78G series, the μA79G is capable of any negative output voltage from its nominal control voltage of −2.23 V to −30 V. Figure 3-33 shows a negative μA79G series regulator with current limiting and short-circuit protection. In this circuit, Q2 protects Q1 from shorts and must have a current rating equal to or in excess of that of the IC regulator.

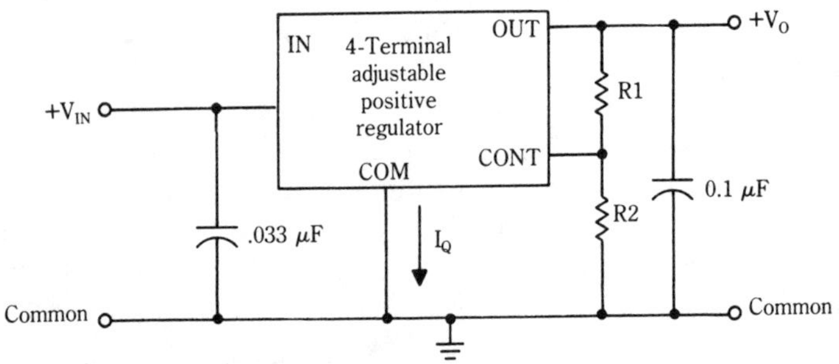

NOTES:

$$V_O = \frac{R1 + R2}{R2} V_{CONTROL}$$

$V_{CONTROL}$ Nominal = 5 V
Recommended R2 current ≈ 1 mA

Fig. 3-30. Basic four-terminal fixed positive regulator. (From the *Fairchild Voltage Regulator Handbook*, copyright by Fairchild; reprinted with permission.)

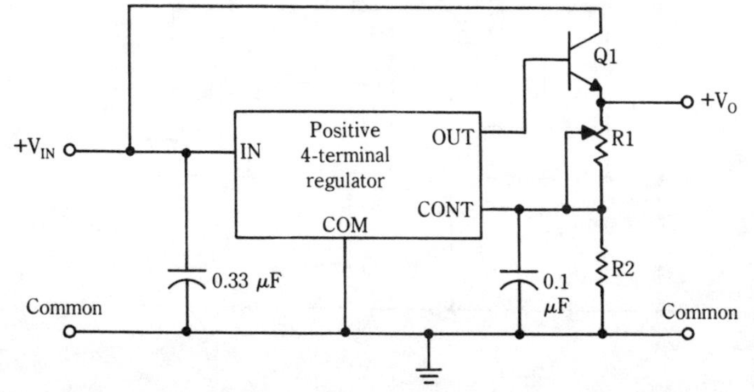

Fig. 3-31. Adjustable positive regulator with increased output current. (From the *Fairchild Voltage Regulator Handbook*, copyright 1978 by Fairchild; reprinted with permission.)

DUAL-POLARITY AND TRACKING SUPPLIES

The simplest dual supply is obtained by using a center-tapped transformer and a positive and negative three-terminal regulator. This is shown in Fig. 3-34. A circuit accomplishing the same result is possible by using two positive or even negative regulators with two isolated transformer windings. This dual-output positive and negative supply is made from two positive three-terminal voltage regulators in Fig. 3-35.

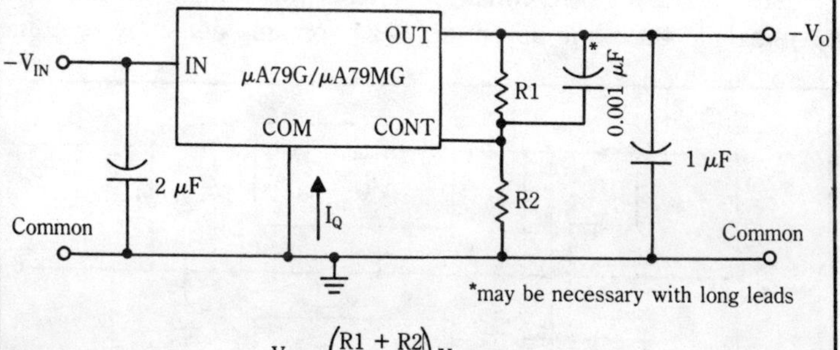

$$V_O = \left(\frac{R1 + R2}{R2}\right) V_{CONTROL}$$

$V_{CONTROL}$ Nominal = – 2.23 V

Fig. 3-32. Basic four-terminal fixed or adjustable negative regulator. (From the *Fairchild Voltage Regulator Handbook*, copyright 1978 by Fairchild; reprinted with permission.)

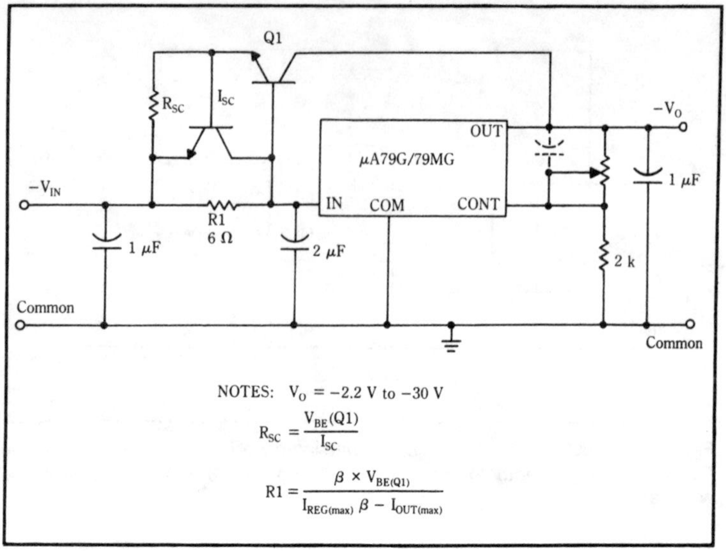

Fig. 3-33. Negative high-current regulator with short-circuit protection (From the *Fairchild Voltage Regulator Handbook*, copyright 1978 by Fairchild; reprinted with permission.)

FIXED AND ADJUSTABLE DUAL-TRACKING REGULATORS

Figure 3-36 shows a fixed dual-tracking regulator for applications requiring tracking between dual supplies. The circuit consists of a positive and negative five-volt regulator, an operational amplifier used as a voltage comparator, and two tightly matched resistors. The voltage comparator tries to maintain its two input terminals at the same potential, as any op amp does. By so doing,

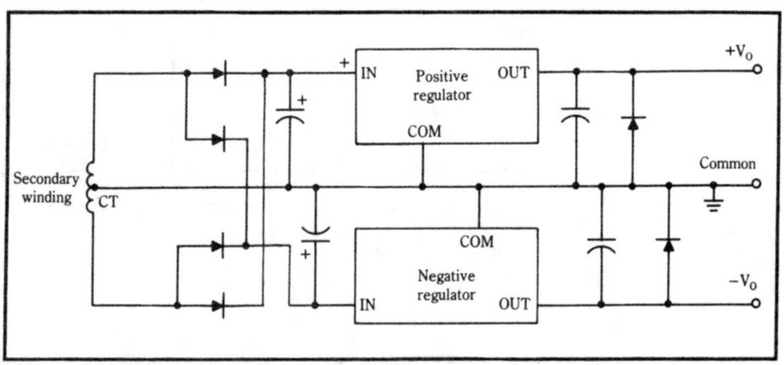

Fig. 3-34. Obtaining dual polarity outputs with two three-terminal voltage regulators. (From the *Fairchild Voltage Regulator Handbook*, copyright 1978 by Fairchild; reprinted with permission.)

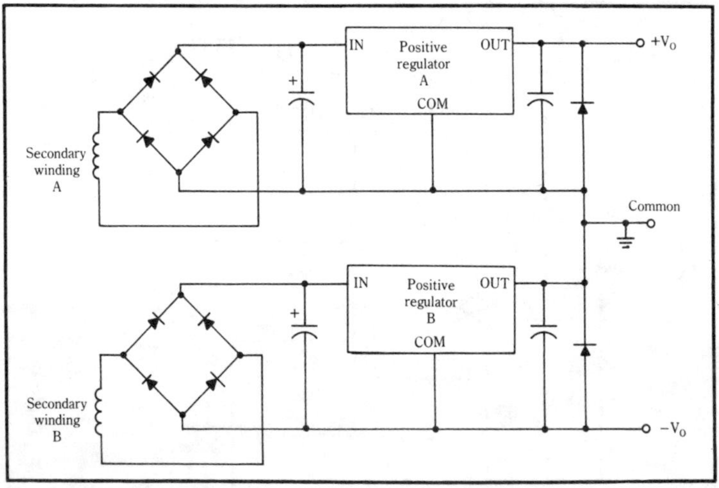

Fig. 3-35. Dual polarity supply with two transformer windings and two positive regulators. (From the *Fairchild Voltage Regulator Handbook*, copyright 1978 by Fairchild; reprinted with permission.)

the noninverting input (+) is referenced to ground so the inverting input (−) is driven to ground. This makes the intersection of R1 and R2 equal to ground. If the point is maintained at ground potential with R1 and R2 being equal, then by a simple voltage divider action, V_0+ and V_0− will be equal in magnitude but opposite in polarity.

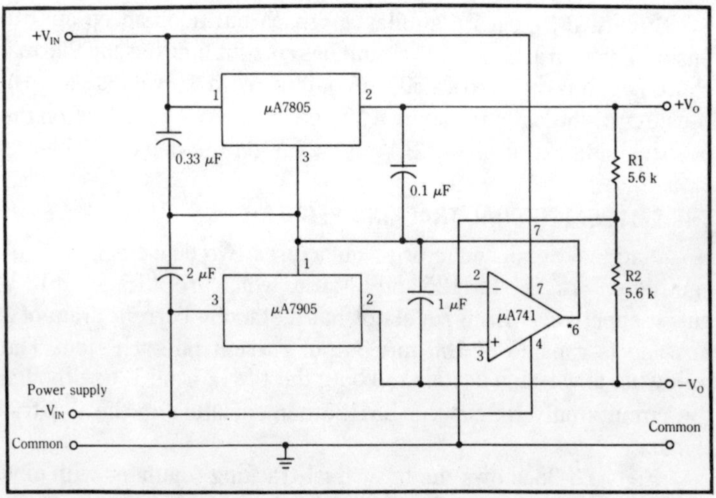

Fig. 3-36. 5V dual-tracking output supply. (From the *Fairchild Voltage Regulator Handbook*, copyright 1978 by Fairchild; reprinted with permission.)

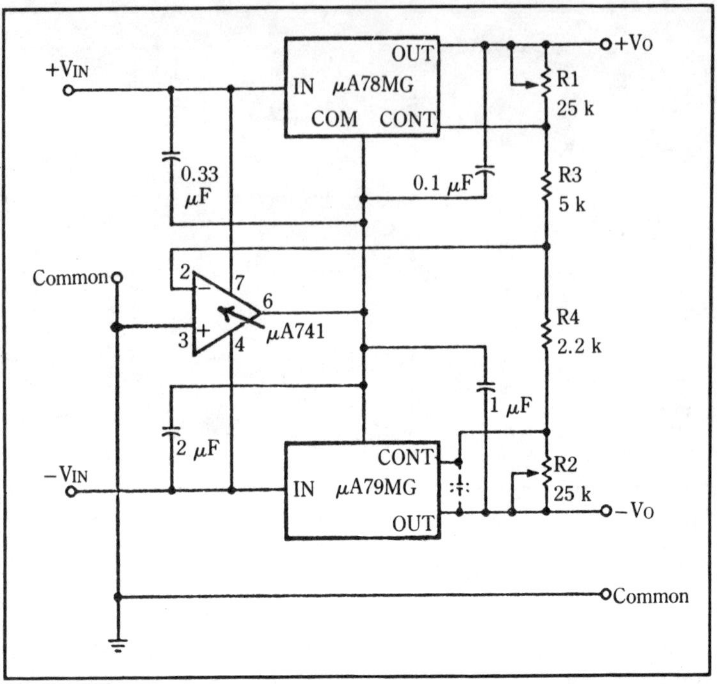

Fig. 3-37. Independently adjustable true dual-tracking power supply. (From the *Fairchild Voltage Regulator Handbook*, copyright 1978 by Fairchild; reprinted with permission.)

Figure 3-37 uses a similar approach but adds an op amp to ensure better tracking. This circuit has current limiting and thermal shutdown. It is capable of a 500 mA output. With the values shown in the circuit, the outputs can be adjusted from +5 V to +30 V on the positive side and from −2.2 V to −27 V on the negative side.

THE LM125/126 DUAL-TRACKING REGULATORS

National Semiconductor manufactures two dual-tracking voltage regulators, the LM125 and LM126, which are ±15 and ±12 V out, respectively. Each regulator has a standby current drain of 3 mA and is capable of 100 mA output current on each side. The following discussion pertains to both the LM125 and 126; although the circuits only reflect the LM125 nomenclature in the illustrations.

Figure 3-38 shows the basic dual-tracking regulator with pins and their functions identified. In its simplest mode of operation, the LM125 merely requires completion of the output current paths from

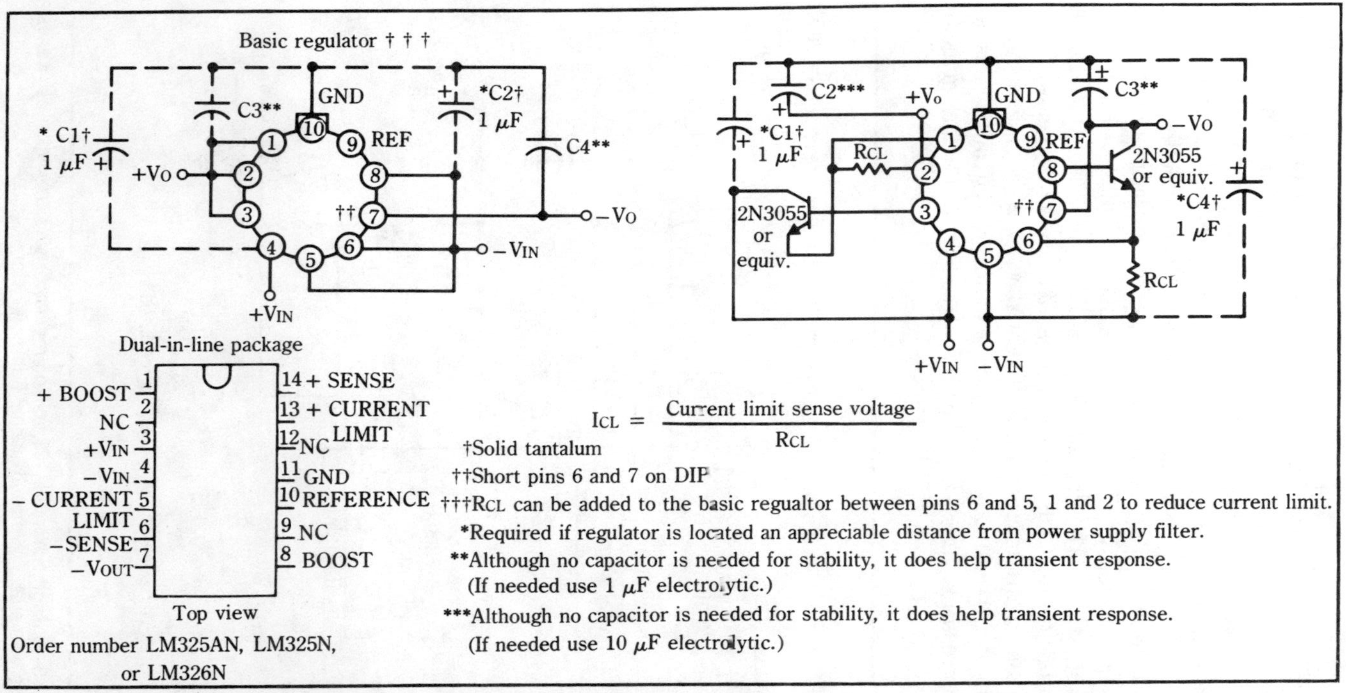

Fig. 3-38. The basic LM125 voltage regulator. (Courtesy National Semiconductor.)

$+R_{CL}$, and $+V_{OUT}$, and $-R_{CL}$ and $-V_{OUT}$. Completion of this current path may be made by directly shorting these pins together. However, if you desire a current limit other than the maximum, you may place resistors between these pins. Figure 3-39 shows a boosted high-current regulator using Darlington pairs for the series-pass elements. This circuit delivers output currents of 10 A at ±15 V with good stability. However, these external Darlington pairs decrease efficiency by increasing the minimum input/output differential to 4.5 V. The normal minimum differential is 2.0 V. During normal operation, the power dissipated by the 2N3772 power transistor is this input-output differential voltage multiplied by the output current; however, upon output short-circuiting this increases to V_{IN} times I_{SC}. The short-circuit current is then given by Equation 3-2.

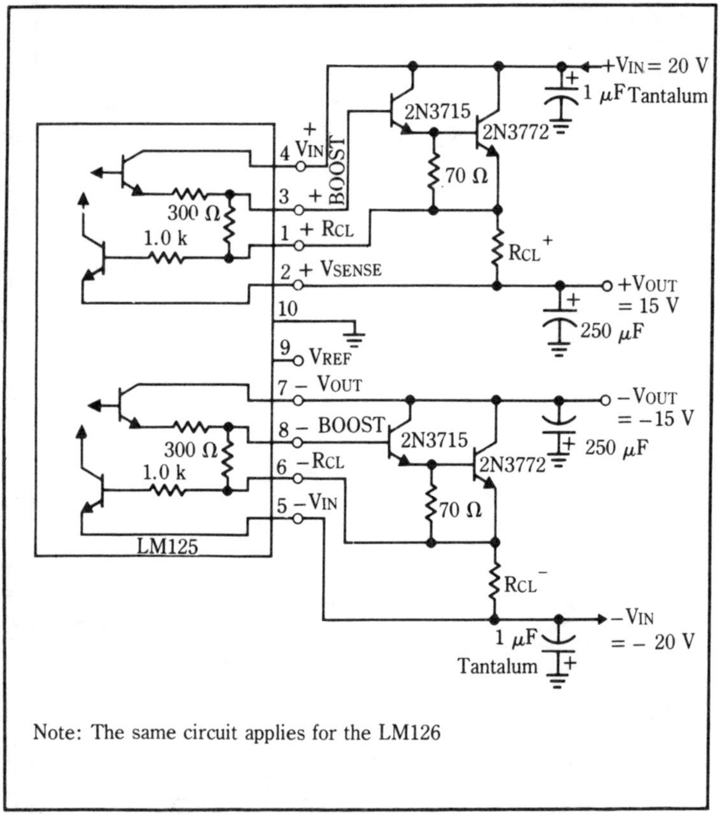

Fig. 3-39. High-current LM125 voltage regulator using Darlington pair series-pass elements. (Courtesy National Semiconductor.)

$$I_{SC} = \frac{P_{MAX}\,(T_c = 25°\,C)}{V_{IN}} \qquad \textbf{Equation 3-2}$$

$$= \frac{150\ W}{20\ V\ (Min)} = 7.5\ A$$

where P_{MAX} = total power dissipated
V_{IN} = input voltage

You can increase the output short-circuit capability to 10 A if you make the short-circuit current less than the load current. A foldback current-limiting arrangement will accomplish this and do so with a no-load to full-load output voltage variation of only 40 mV.

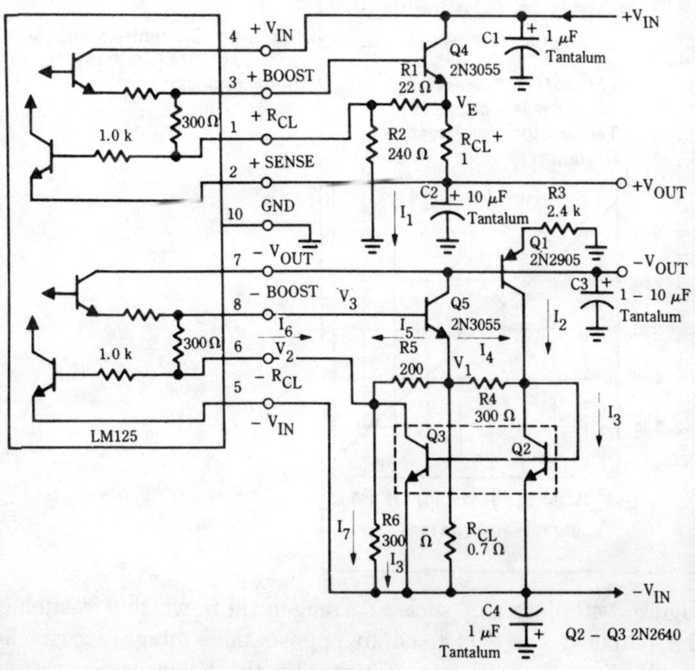

Note: For LM 126; $V_{IN}+ = 25$ V, $V_{IN} = 25$V
$R_1 = 20$, $R_2 = 180$
$R_{CL} + = 0.9$
$R_3 = 1.35$ K, $R_6 = 290$; $R_{CL} - = 0.9$
$I_{FB} = \pm 2A$, $I_{SC} = \pm 0.75A$

Fig. 3-40. Foldback current limiting circuit. (Courtesy National Semiconductor.)

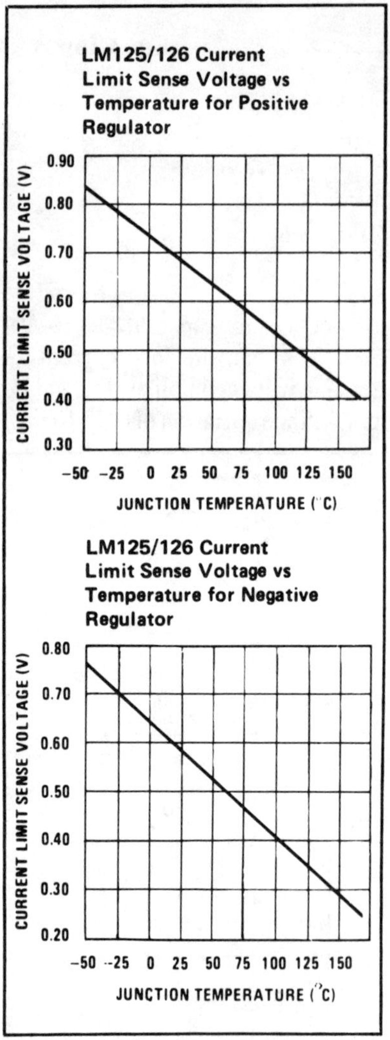

Fig. 3-41. Current-limiting sense voltage graph. (Courtesy National Semiconductor.)

Figure 3-40 illustrates such an arrangement in which a fraction of the output voltage is used to oppose the voltage across the current-limit sense resistor. Current limiting begins to occur when voltage across the sense resistor is higher than this opposing voltage by the amount shown in Fig. 3-41. If the output is grounded, this opposing voltage is no longer present so current limiting occurs at a lower level. The circuit in Fig. 3-40 uses a programmable current source to provide a constant voltage drop across R5 for the negative portion of the regulator. The positive portion of the regulator uses a simple voltage divider. The positive foldback current-limiting circuitry uses R1 and R2 as a voltage divider with this divided-down voltage opposite in polarity to the voltage drop across R_{CL} +. When the load current increases to where the IR drop across R_{CL} + is equal to the current-limit sense voltage (refer to Fig. 3-40) plus the drop across R1, the voltage regulator begins to limit current.

It is still possible, even with this 10 A beast, to electronically shut it down. This can be done by several methods. One is by inserting a relay or saturated bipolar device in series with the inputs or outputs. If you desire a little more sophistication than this brute-force method, then you may wish to electronically short the internal voltage reference to ground.

This will prove effective with only one slight problem—the outputs will not absolutely go to ground. The positive output will remain at +700 mV while the negative output will remain at −300 mV. The circuit in Fig. 3-42 accomplishes electronic shutdown with

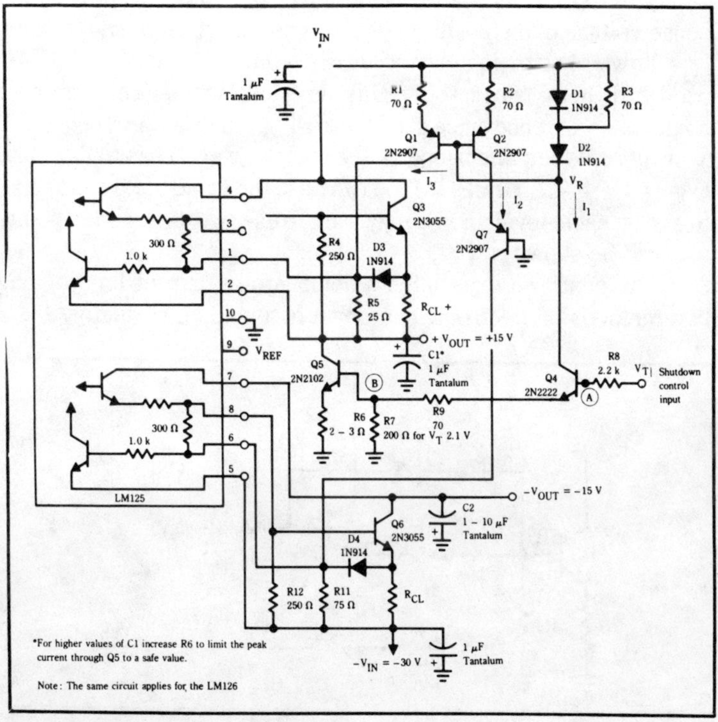

Fig. 3-42. Electronic shutdown with a boosted LM125 voltage regulator circuit. (Courtesy National Semiconductor.)

each output going fully to ground. The shutdown signal, as shown, is for a TTL logic level; however, by using the following relationship any voltage above approximately 1.5 V or $2 \times V_{BE}$ drops can shut the regulator down. Note how the values of R8 and R9 can be adjusted to accomplish this:

$$V_T \simeq \left(\frac{R8}{R3\beta_{Q4}} + \frac{R9}{R3}\right) V_{BE} + 2\,V_{BE}$$ **Equation 3-3**

where V_T = threshold shutdown voltage
V_{BE} = base-to-emitter voltage for Q_4
B_{Q4} = Q_4 current gain
R3 = 70 Ω
R8, R9 = adjustable resistors

If transistors Q1 and Q2 are closely matched and R1 = R2 = R3, then simultaneous equal shutdowns occur. The current I_3 causes a voltage drop across R5 which either equals or exceeds the limit sense voltage of the positive regulator and causes it to shut down.

Power dissipation may occur when the design of $V_T = 2.0$ V is used and the V_T goes above this threshold value. This can be a common occurrence since a TTL logic high can vary as much as 1.2 V or more. When this happens, the transistor (Q4) has to dissipate $(V_{IN} - 3\,V_{BE} - V_T)$ times I_1. You can raise the value of R8 and keep I_1 to a reasonable level, or you may want to use two diodes from points A and B as shown in Fig. 3-43.

The next two experiments demonstrate current limiting and two methods of electronic shutdown of a voltage regulator.

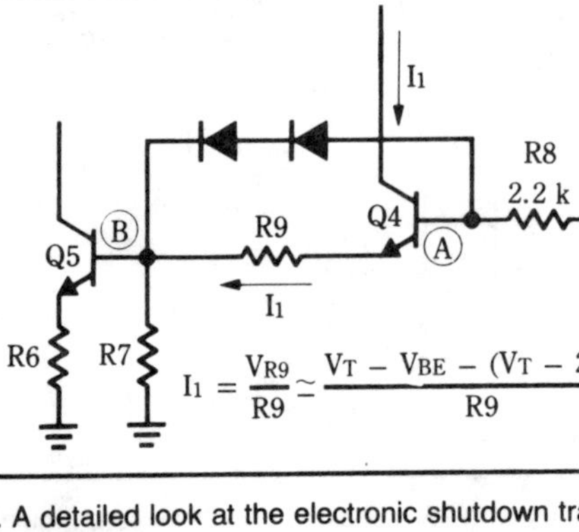

$$I_1 = \frac{V_{R9}}{R9} \simeq \frac{V_T - V_{BE} - (V_T - 2\,V_{BE})}{R9} = \frac{V_{BE}}{R9}$$

Fig. 3-43. A detailed look at the electronic shutdown transistor. (Courtesy National Semiconductor.)

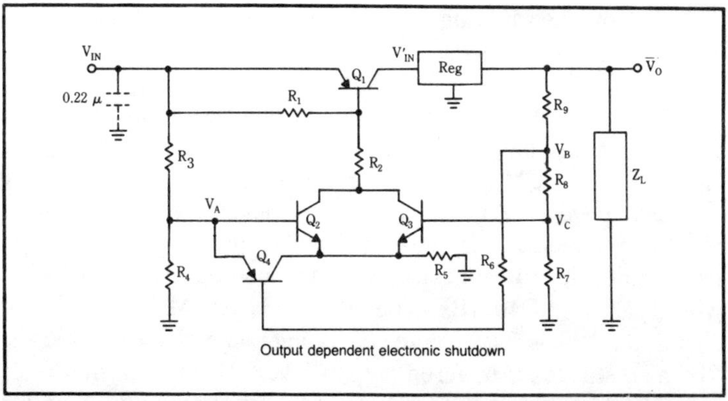

Fig. 3-44. Circuit for Experiment 3-1. (Courtesy National Semiconductor.)

EXPERIMENT 3-1

This experiment demonstrates an output-dependent shutdown rather than an electronic shutdown from a logical level.

When the power is first applied to the circuit in Fig. 3-44, Q2 turns on and saturates Q1, the regulator's output ramps up to turn on Q2, thus maintaining Q1 in the ON state. When the output is shorted or greater than 1 A of current is drawn, Q3 turns OFF. Q4 is turned on to clamp Q2 OFF. Q1 loses base drive and enables Q1 to restart the regulator. Q4 may be any small signal PNP polarity transistor and Q2 and Q3 do not have to be matched at all, unlike our experiments in Chapter 2. Some suggested values are given in Table 3-4. Use a 24 V regulator as specified because these component values have been calculated for a 24 V regulator.

1. Build the circuit in Fig. 3-44.
2. Short the output.
3. Let the circuit draw various levels of load current around 1 A.

Table 3-4. Parts List for Experiment 3-1.

	LM340K-15	
V_{IN} = 25 V	Q_1 = TIP32	R1 = 500Ω
V_0 = 15 V	Q_2 = 2N4141	R2 = 250 Ω, 2 W
I_0 = 1 A	Q_3 = 2N4141	R3 = 3 kΩ
V_A = 2.5 V	Q_4 = 2N2906	R4 = 330 Ω
V_B = 8 V		R5 = 62Ω
V_C = 4.6 V		R6 = 2 kΩ
		R7 = 1 kΩ
		R8 = 680 Ω
		R9 = 1.5 kΩ

4. Note the variation in the shutdown point with respect to the heating of the transistors.

Note

A number of circuits presented in this text as well as in the experiments use logic or digital integrated circuits. A very popular family of logic is CMOS and the part numbers could become baffling if not explained. RCA is one of the largest and oldest manufacturers of CMOS integrated circuits so I will explain their numbering convention which is almost universally adopted industry wide. A typical part is a CD4011BE. The "C" stands for CMOS, the "D" for digital, the "B" is for a voltage operating range of 3 V to 20 V; the "B" also stands for buffered outputs. An "A" suffix in this place indicates nonbuffered outputs and a more limited voltage operation range of 3 V to 15 V. The "E" stands for a plastic case and nonmilitary temperature range. An "F" in this place indicates a military temperature range and a ceramic DIP package.

EXPERIMENT 3-2

This experiment demonstrates a circuit with TTL logic level shutdown characteristics.

1. Build the circuit shown in Fig. 3-45.
2. Apply a TTL logic level high and note what happens.
3. Remove the high and ground the input to the base of Q2.
4. Make R3 larger and larger until transistor Q2 will no longer act as a switch. You may also manipulate the value of R2 by using the following equation:

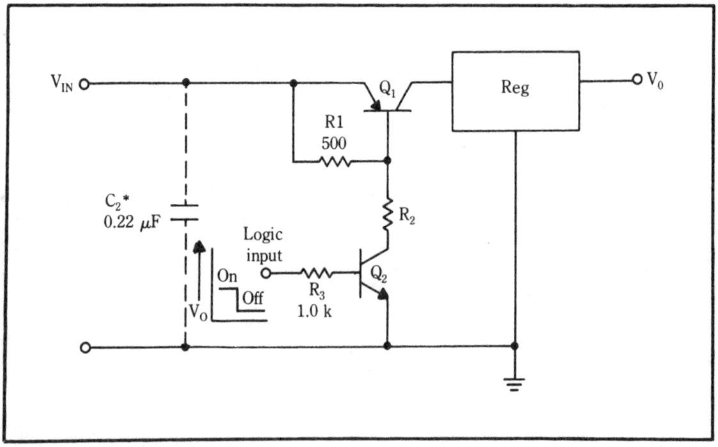

Fig. 3-45. Circuit for Experiment 3-2. (Courtesy National Semiconductor.)

$$R2 = \frac{V_{IN} - .1V}{I_{OUT}} (\beta_{sat} \text{ of } Q1)$$

The .1V is the voltage from the collector to the emitter of a conducting transistor. β_{sat} equals the current gain of Q1 when in saturation.

Chapter 4

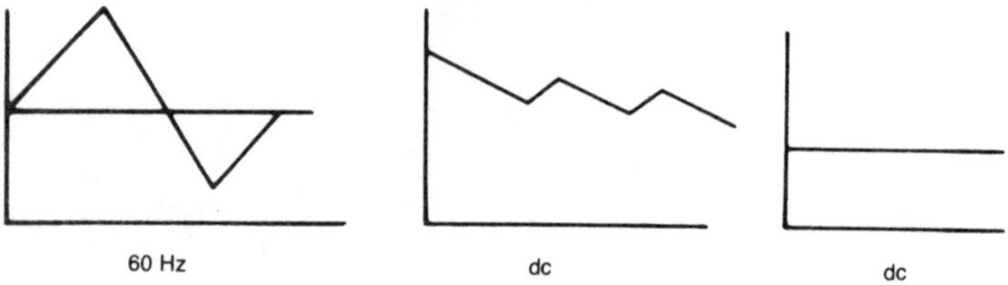

Switching Power Supplies

Up to this point, only linear voltage regulators have been discussed, but what is the difference between linear and switching voltage regulation? A linear voltage regulator has its series-pass transistor operating somewhere between saturation and cutoff, in the linear region. A switching regulator has its series-pass transistor operating either at saturation or cutoff. The "switching" voltage regulator's series-pass transistor essentially acts as a switch which opens and then closes, and hence derives its name.

The difference in the way these two voltage regulators' series-output transistors work is very important. This is also why the switcher is so much more efficient. A transistor operating in saturation or cutoff wastes very little power. In saturation, the transistor has very little voltage across it but a large current flowing through it. In cutoff, the transistor has very little current flowing but the full voltage across it. In either case, the transistor is operating at or near maximum efficiency.

The following is a list of the advantages and disadvantages of this arrangement.

Advantages

- Efficiency. Up to 93%, while a linear is typically 45% efficient.
- Versatility. A given input can be converted up or down in voltage or have its polarity reversed.
- Size and Weight. Switches can be made smaller and lighter.

Disadvantages

● Complexity. They are more complex and require additional components.

● Noise and Ripple. Switchers generate more electrical noise and ripple than a linear.

● Response. The output cannot respond to transients or very sudden changes in load condition requirements as well as a linear.

SWITCHING REGULATOR CONCEPTS

A switching regulator stores energy in inductors and capacitors (which do not consume power) and supplies energy to the load as required. The switcher cycles on and off likewise turning the series-pass transistor on and off with a controllable "duty cycle" (time on versus time off percentage ratio) doling out energy to the load as required. Refer to Fig. 4-1 for a block diagram of a switching voltage regulator.

The overall block diagram will be briefly examined with an explanation of each one of the blocks within the block diagram. This will be followed by a detailed examination of three very popular switching voltage regulator integrated circuits.

The switching power supply is a modernized version of the old electromechanical vibrator which was used extensively to supply the 100 V plus potentials required for tubes in car radios. The modern switcher with all of the regulation and control circuitry on a single integrated circuit is an immensely refined version of this electromechanical vibrator. Refer to Fig. 4-1 and note that the rectifier converts ac voltage into dc and smooths it with a filter

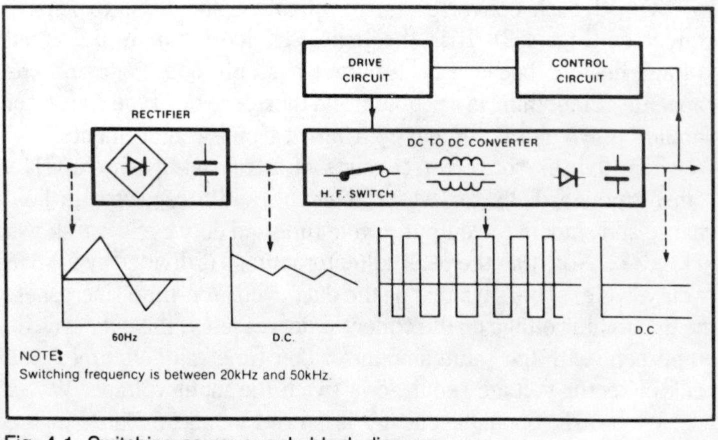

Fig. 4-1. Switching power supply block diagram.

capacitor. This dc is used to power an electronic chopper (this replaced the electromechanical vibrator) which operates typically in the frequency ranges of 20 kHz to 50 kHz. The chopped dc is applied to the primary of a transformer with its secondary rectified and smoothed to give the required dc output. The purpose of the transformer is naturally for converting one voltage to another, but additionally it provides isolation or negates and minimizes the influence the input has on the output. The output voltage is sensed by the control circuitry and provides a signal (usually a dc level) that is a correcting signal fed back to adjust the duty cycle of the switching transistor. This is accomplished via the drive circuitry and keeps the output voltage constant and therefore provides voltage regulation. If the input rectifier section were to be deleted, a battery or dc supplying circuit could run this switcher equally as well.

The rectifier and filter capacitor are specified keeping in mind a huge possible inrush current caused when the output transistor is switched on and the output responds rapidly. However, there are slow-start circuits to help prevent this. At higher frequencies, 200 kHz to 1 MHz, a whole fresh new look is required for component selection as will be shown later.

DC-TO-DC CONVERTERS

Depending on the requirements, the dc-to-dc converter may be one of the three different types.

The Flyback Converter

The flyback converter is the simplest and least expensive converter (Fig. 4-2). It is the logical choice for multiple output voltage designs because each output has only one diode and one capacitor. Smoothing is a problem and therefore this type converter should be avoided if you have stringent ripple requirements.

The flyback converter consists of a transistor switch, Q1, a transformer, a diode to "steer" current, and a capacitor to hold charge and smooth the output. Cycle time and duty cycle are shown in Fig. 4-2. Note that the peak collector voltage is divided by $1-\delta$ (δ = duty cycle). This says that as the duty cycle "on time" increases, the maximum voltage on the collector decreases. If the duty cycle is 50 percent with the same amount of "on time" as "off time," the peak collector voltage required is twice the input voltage, V_i.

When Q1 conducts, energy is stored or accumulated in the

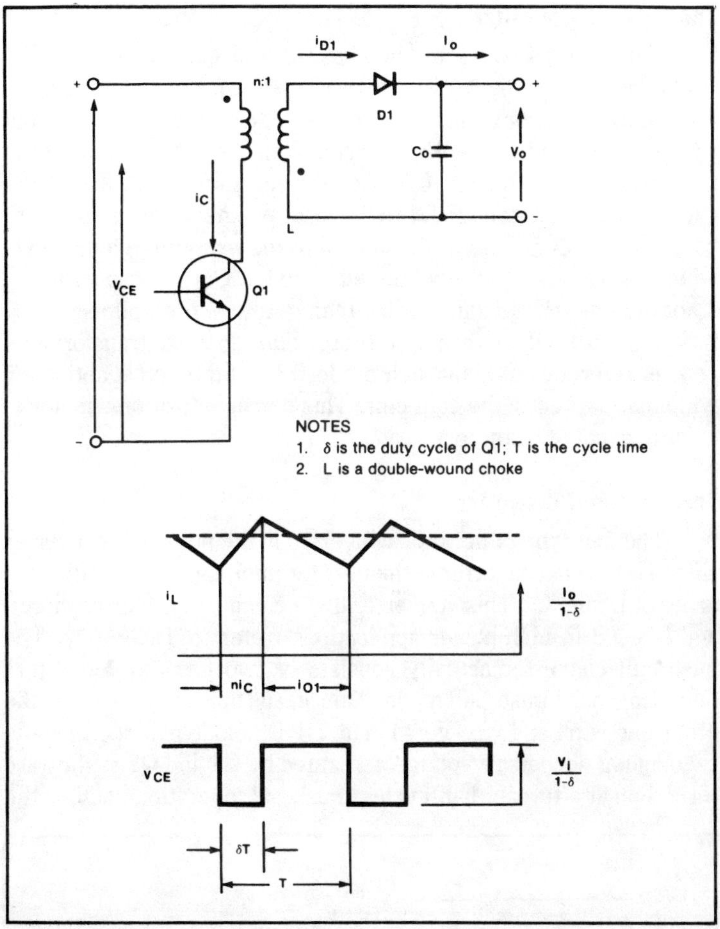

Fig. 4-2. A flyback dc-to-dc converter.

magnetic field and while the inductor's current rises the diode is reverse biased. This is followed by a discharging of energy into the output capacitor and the load during the "flyback" period. This "flyback" period is when Q1 is off and the current in the inductor is collapsing or falling and the diode is forward-biased and conducting. During conduction, Q1 continues delivering charge to the load, thus providing a smooth, continuous supply of power.

This scheme is very simple because only one transformer or inductive element is required as opposed to the following two converters which require two inductive elements (either transformers or chokes).

The Forward Converter

The forward converter in Fig. 4-3 is useful for low voltage applications with more stringent ripple requirements than the flyback dc-to-dc converter can provide. The high-frequency output ripple is limited by a choke in series with the output. During the time the transistor switch Q1, is on, the energy is stored in both the choke and passed through D1 to the load. When the transistor is off, part of the energy is still transferred to the load through diode D2 which is called a "freewheeling" diode. Output capacitor C_o smooths the ripple due to switching transistor spikes. After switching turn-off the magnetic energy built up in the transformer's core is returned to dc through diode D3 and the demagnetization winding attached to the collector. This demagnetization winding is closely coupled to the primary.

The Push-Pull Converter

The last type of dc-to-dc converter is the push-pull converter (Fig. 4-4). This converter is the best for minimum ripple with a set value of L_o and C_o. This converter has a complex base drive circuit and is used in high-power applications (refer to Table 4-1). The push-pull converter actually consists of two forward converters operating in the push-pull mode. This means that one is on while the other one is off and vice-versa. The D1-D2 diode pair rectifies the rectangular secondary voltage generated by Q1 and Q2 as they are turned on at alternate half-cycles. Push-pull operation doubles the

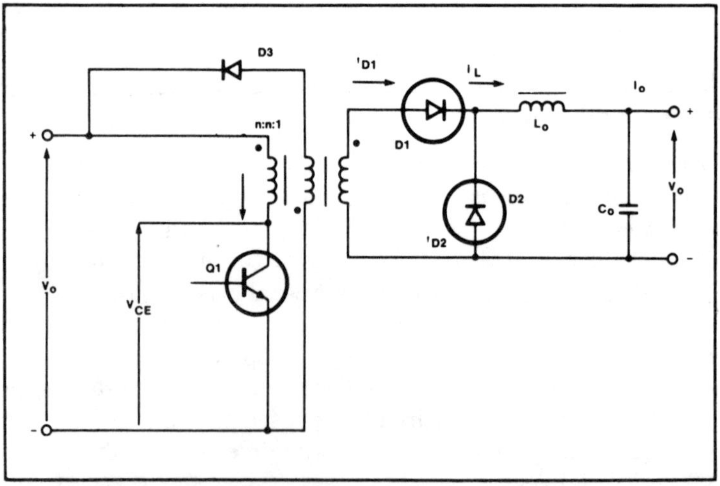

Fig. 4-3. A forward dc-to-dc converter.

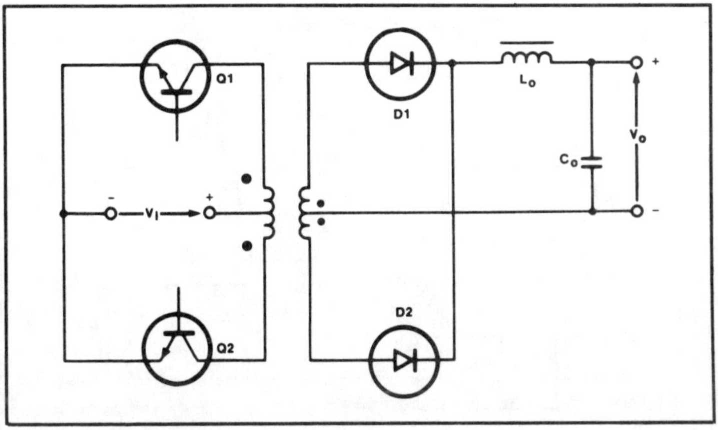

Fig. 4-4. A push-pull dc-to-dc converter.

Table 4-1. Dc-to-Dc Converter Comparison. (From *Design of Solid-State Power Supplies* 2nd Ed., by Hnatek, Eugene R., published by Van Nostrand Reinhold company; reprinted with permission.)

Type	Advantages	Disadvantages
Flyback	• Isolation from supply to line • Output doesn't need to be filtered by choke • Simple circuitry • Low component count • Easily provides multiple outputs • Simple power transistor base drive circuitry • Doesn't require output transformer • Useful in high voltage multiple output applications	• High output ripple • Requires large core inductors • Low power output levels
Forward (single ended or ringing choke)	• Low output voltage ripple • Easily provides multiple outputs • Needs only single switching transistor • Simple power transistor base drive circuitry • Useful in low power applications	• Requires addition of transformer for line isolation • Extra winding and components required for multiple outputs
Push-pull	• Low output voltage ripple • Good regulation • Uses small transformer component cores • Provides automatic line isolation • High power output capability • Useful in high power high performance single output applications	• High component count • Complex circuitry • Complex power transistor base drive circuitry • Subject to dc imbalance which can cause core saturation • Extra winding and components required for multiple outputs

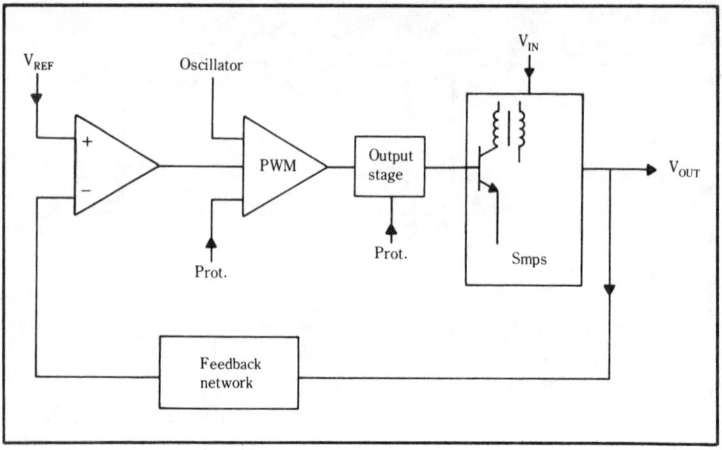

Fig. 4-5. A switching voltage regulator block diagram.

frequency of the output ripple which makes it easier to filter, just as was the case with non-half-wave rectifier schemes discussed previously.

You may have noted in the three previous explanations that it was assumed the base drive was present and functioning properly to keep the transistor turned on for just the right amount of time, thus maintaining proper regulation. The bases on all of those transistors were conspicuously blank or devoid of drive circuitry.

Look at Fig. 4-5 and observe the control loop that is responsible for sensing the output voltage, feeding it back around to an error amplifier, and then comparing this voltage to a reference voltage. This is what controls the drive to the base. Up to this point, all of this is very similar to a linear voltage regulator. However, note the addition of a PWM (pulse-width modulator) and its output drive stages. The PWM is a pulse-width modulator with three inputs. The error amplifier is one input with the second being an oscillator generating either a sawtooth or triangle wave. The third input to the PWM is one for protection and can be for overvoltage, overcurrent, etc. As the pulse width is varied so is the duty cycle or the amount of time the power transistor is turned on. By varying this, the amount of energy taken from the input voltage will also vary and it is this mechanism which is responsible for stabilizing the output against line and load variations.

OTHER TYPES OF REGULATORS USED IN SWITCHERS

The pulse-width modulator regulator just described is the most

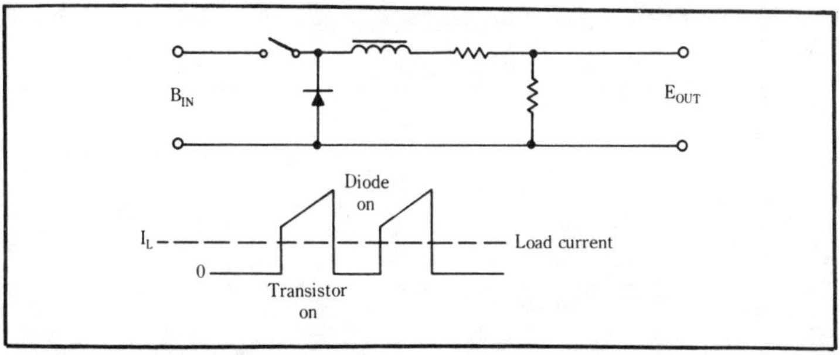

Fig. 4-6. The buck regulator.

common type but there are at least four other regulator types in existence.

The buck regulator takes power from a primary voltage source such as a fuel cell or battery and does so in spurts with the duration of these bursts depending on the line voltage. The dc output is the average of these pulses. Refer to Fig. 4-6. The boost regulator boosts the input voltage to the desired level which is greater than the input voltage. Refer to Fig. 4-7.

The phase-control regulator has two square waves that are added together either in-phase or out-of-phase. When added together in-phase, the maximum output voltage is obtained. Actually, in this configuration any output voltage can be obtained by adjusting the turns ratio of the transformer. Refer to Fig. 4-8.

The buck-boost regulator stores energy in a choke or inductor. When the switch is opened, the choke discharges—the faster the discharge the higher the voltage. Refer to Fig. 4-9.

NE5560

At this point, we could delve quite deeply into an analysis of all of this control circuitry driving the base or bases of the output transistor or transistors in the three types of dc-to-dc converters we have studied, but this is needless. All of this control circuitry is now

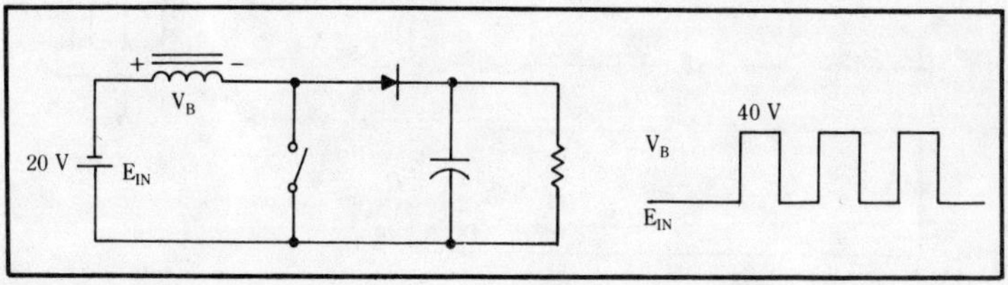

Fig. 4-7. The boost regulator.

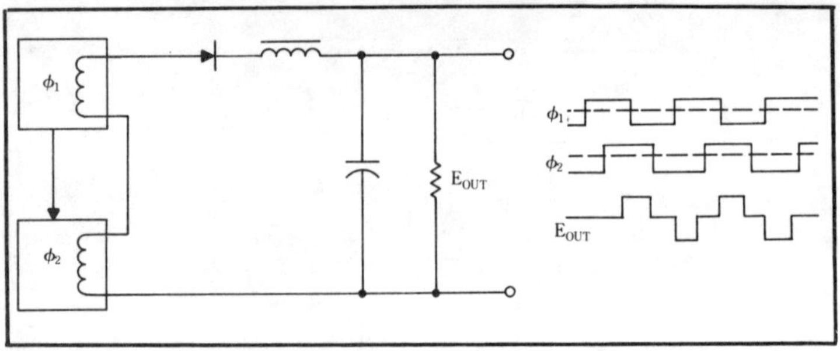

Fig. 4-8. The phase-control regulator.

contained on several single integrated circuits. Let's first examine two very popular ones made by Signetics and then one made by Fairchild.

Signetics' NE/SE5560 switched-mode power supply control circuit IC is shown in Fig. 4-10. The "NE" prefix indicates a commercial temperature range with the "SE" prefix indicating a military component. The 5560 is a 16-pin DIP with internal temperature-compensated voltage reference, zener reference, sawtooth generator, pulse-width modulator (PWM), output stage, and various protection circuits.

Before explaining the detailed block diagram, two terms must be defined. The first is *soft start* which is the gradual increase of the transistor duty cycle after switch-on, causing a slow rise of the output voltage. This prevents an excessive inrush current due to a capacitive load or charging of the output capacitor. The second term is *synchronization*. Synchronization is having all switching power supplies switch in unison at the same frequency and is used to prevent interference due to the difference in free-running frequencies (for example, in a system where a low-power switching power supply supplies the base drive circuit of the output switching tran-

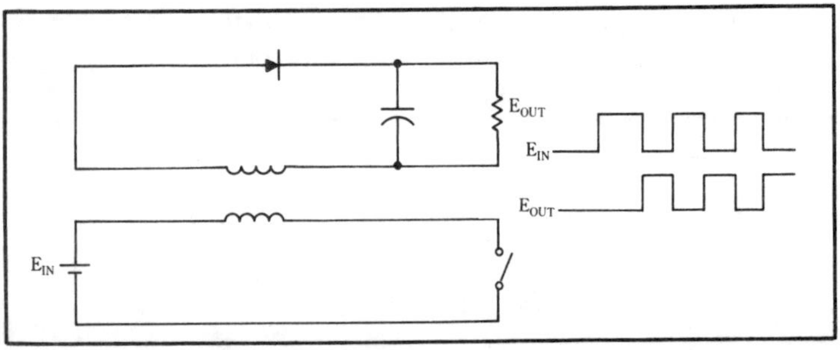

Fig. 4-9. The buck-boost regulator.

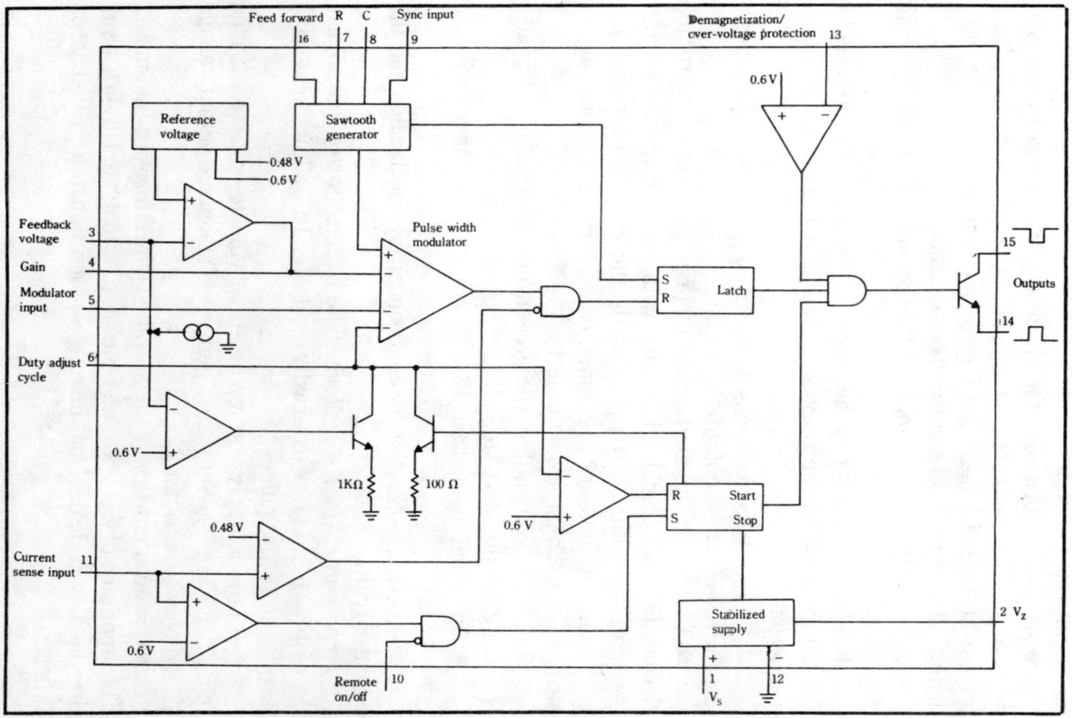

Fig. 4-10. An NE5560 block diagram.

sistor in a high-power switching power supply). A brief description of each pin follows, then we will discuss the overall operation of the IC. Refer to Fig. 4-10, the 5560 block diagram and note the following features:

- A temperature-compensated reference source.
- An error amplifier with pin 3 as input. The output is connected to pin 4 so that the gain is adjustable with external resistors.
- A sawtooth generator with a TTL-compatible synchronization input (pins 7, 8, and 9).
- A pulse-width modulator with a duty-cycle range from 0 to 95 percent. The PWM has two additional inputs: pin 6 can be used for a precise setting of δ max. Pin 5 gives a direct access to the modulator, allowing for constant-current operation.
- A gate at the output of the PWM provides a simple, dynamic current limit.
- A latch that is set by the flyback of the sawtooth and reset by the output pulse of the previously mentioned gate and prohibits double pulsing.
- Another latch functions as a start-stop circuit; it provides a fast switch-off and a soft start.
- A current protection circuit that operates via the start-stop circuit. This is a combined function with the current-limit circuit; therefore, pin 11 has two trip-on levels, the lower one for cycle-by-cycle current limiting, the upper one for current protection by means of switch-off and soft start.
- A TTL compatible remote on/off input at pin 10, also operating via the start-stop circuit.
- An inhibit input at pin 13. The output pulse can be inhibited immediately.
- An output gate that is commanded by the latches and the inhibit circuit.
- An output transistor of which both the collector (pin 15) and the emitter (pin 14) are externally available. This allows for normal or inverse output pulses.
- A power supply that can be either voltage or current driven (pins 1 and 12). The internally-generated stabilized output voltage V_z is connected to pin 2.
- A special function is the so-called "feed forward" at pin 16. The amplitude of the sawtooth generator is modulated in such a way that the duty cycle becomes inversely proportional to the voltage on this pin:

$$\frac{C}{V_{16}} \approx \delta$$

● Loop fault protection circuits assure that the duty-cycle is reduced to zero or a low value for open or short-circuited feedback loops.

The 5560 uses a series linear voltage regulator and has a stable 8.5 V output. The 5560 has its zener voltage (V_Z), pin 2, capable of sinking 5 mA and is used for the precise setting of δ max. To review the δ max concept again, δ is the transistor's duty cycle and is defined as $V_{CE} = \frac{V_{IN}}{1-\delta}$. The stabilized power supply, as mentioned previously, can be driven from a dc voltage source of between 10.5 V and 18 V or by a current source at pin 1 via a series limiting resistor. The stabilized power supply within the 5560 limits the current inputs to typically 23 V for 10 mA and 30 V for 30 mA. When the supply voltage falls below this 10.5 V limit, the output pulses are inhibited. When the 10.5 V limit is again reached, the output pulses recommence but are brought up by the soft-start circuitry. The 5560 IC is very efficient, drawing only 10 mA or less at 12 V.

The sawtooth generator within the block diagram is shown in more detail in Fig. 4-11. When a resistor is placed between pins 7 and 12 (ground), a linear increasing voltage appears at pin 8 until it reaches the comparator's trip point of 5.6 V, at which time the RS

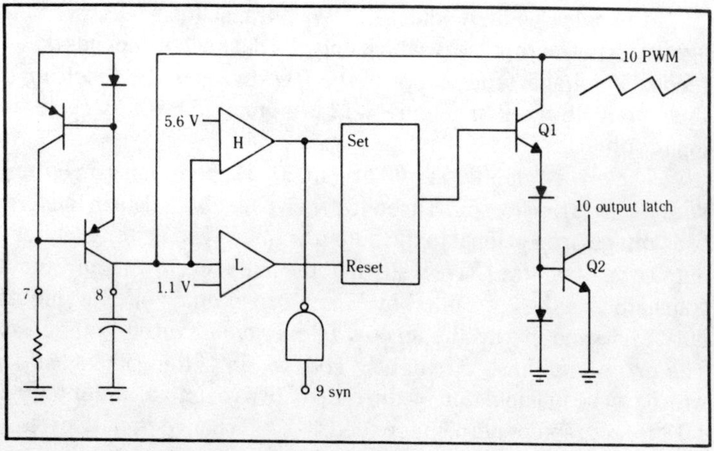

Fig. 4-11. The NE5560 sawtooth generator.

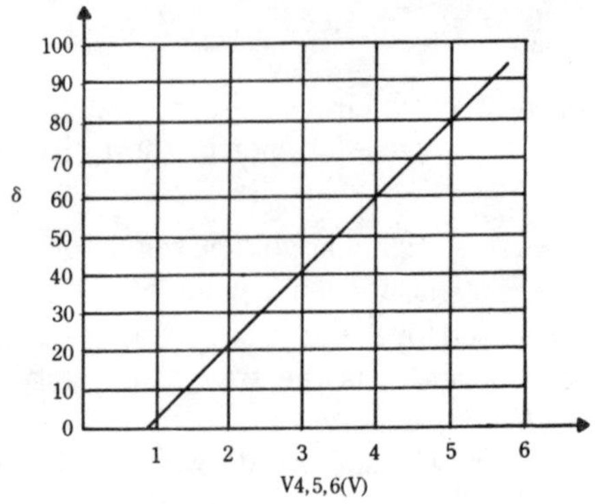

Fig. 4-12. PWM duty cycle vs input voltage.

flip-flop is set and Q1 discharges (C8 toward ground) until it reaches 1.1 V, where comparator L resets the RS flip-flop. During this "flyback" period, Q2 inhibits any output pulses. If the voltage on pin 9 is maintained at a level above 2 V, the setting of the sawtooth generator is prohibited and this allows synchronization at a frequency lower than the free-running rate.

The reference voltage source is an ultrastable bandgap reference of 3.72 V with a temperature coefficient of less than 100 ppm/° C.

The pulse-width modulator (PWM) translates a feedback voltage into a periodic pulse of which only the duty cycle depends on the feedback voltage. The output of the PWM causes the resetting of the output RS flip-flop. Figure 4-12 is a graph of the duty cycle vs. input voltage.

The error amplifier is a 60 dB (1,000 : 1) open-loop gain op amp with its inverting (−) pin 3 used to receive feedback data in the form of a voltage proportional to the output voltage. The error amplifier's output goes to the PWM so that the duty cycle of the output transistor can be controlled by being turned on or off, varying its duty cycle and in turn the amount of energy delivered to the load. The error amplifier's output also goes to pin 4, the gain adjust pin which can be manipulated by the ratio of two resistors. Refer to Fig. 4-13 for a graph of percentage feedback vs. change in duty cycle.

In a system with several power supplies, it is often necessary to switch them on in a sequential manner. Also, as shown previously

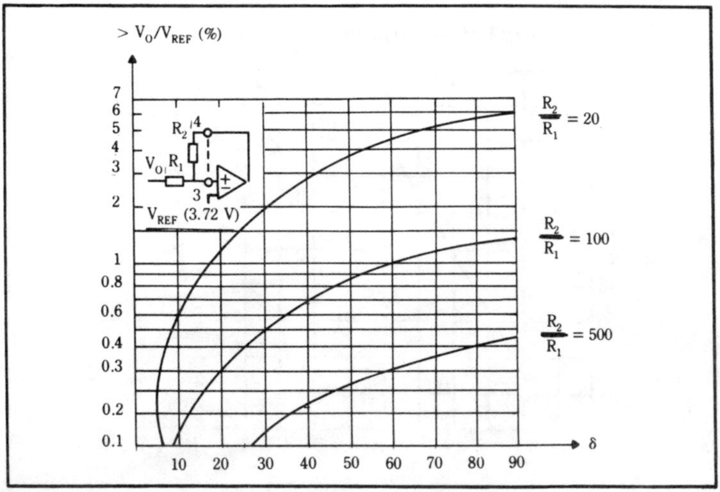

Fig. 4-13. Percentage feedback vs change in duty cycle. (From Signetics data sheets, copyright 1980 by Signetics; reprinted with permission.)

and as will be shown in the next chapter on the 723 voltage regulator, it is often desirable to have a logic level such as a TTL level shut off or turn on a power supply. The 5560 has these provisions. Pin 10 (TTL Remote On-Off) inhibits the power supply for levels below 0.8 V and enables the power supply for levels above 2.0 V. If

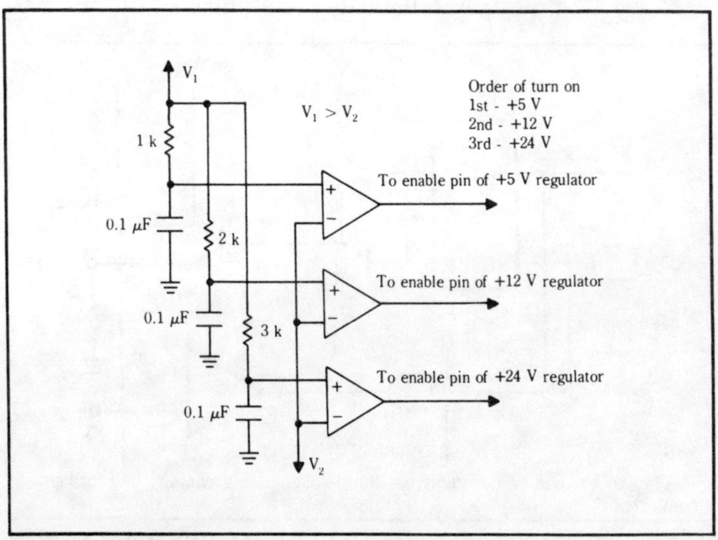

Fig. 4-14. A power-up sequencer circuit. (From Signetics data sheets, copyright 1980 by Signetics; reprinted with permission.)

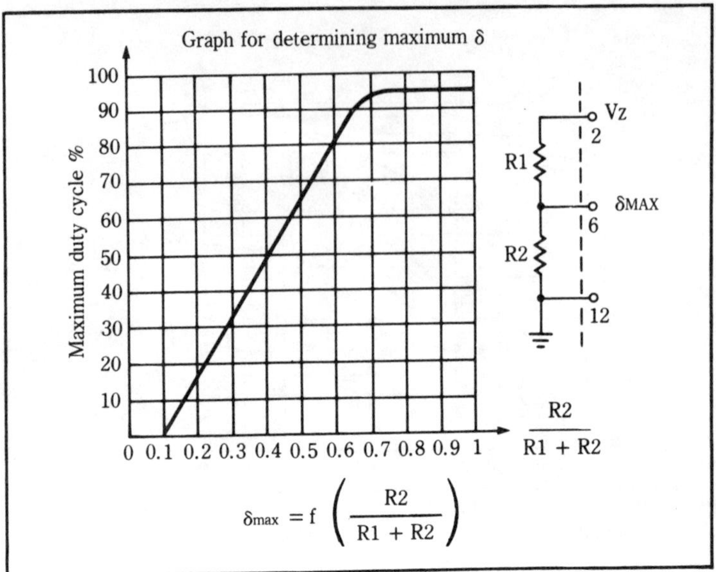

Fig. 4-15. Ratio R1, R2 for duty cycle. (From Signetics data sheets, copyright 1980 by Signetics; reprinted with permission.)

three power supplies were desired to be turned on in sequence, the circuit in Fig. 4-14 might be used.

Figure 4-15 is a graph of how the ratio of two resistors on pins 2, 6, and 12 can establish the duty cycle. Figure 4-16 shows the

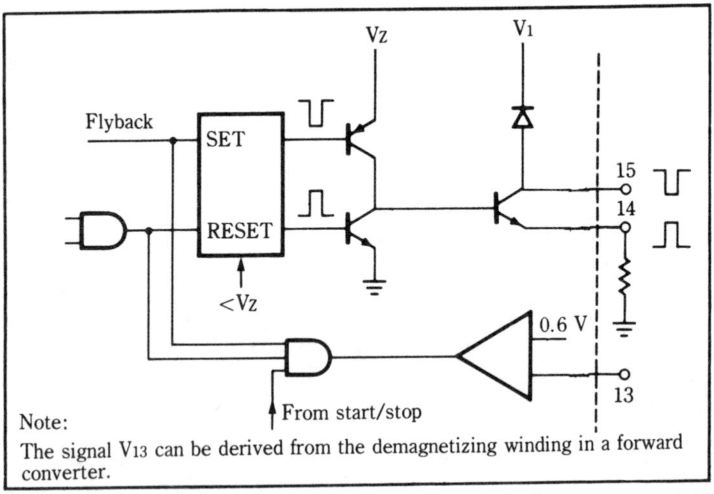

Fig. 4-16. The NE5560 output stage. (From Signetics data sheets, copyright 1980 by Signetics; reprinted with permission.)

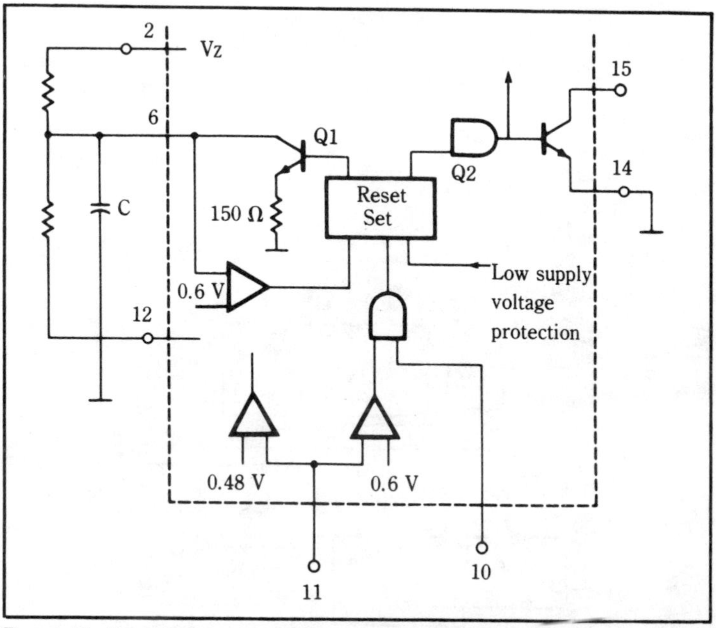

Fig. 4-17. The NE5560 stop/start circuitry. (From Signetics data sheets, copyright 1980 by Signetics; reprinted with permission.)

output stage which has a 40 mA maximum output capability with a +5 V maximum voltage at the emitter. Note how the base of the output transistor can be clamped by either a set or reset pulse or a command from the stop-start circuitry. In any event, any of these will turn the output transistor off by short-circuiting its base.

The stop-start circuitry, as shown in Fig. 4-17, is for protecting the IC and its associated circuitry. This is accomplished by stopping the output pulses for several periods when a fault occurs and keeping them stopped for several periods. Upon start-up the output pulses are again small and have their duty cycle gradually increased. If the fault persists this cycle starts over and keeps repeating until the fault goes away.

Figure 4-18 shows a 5560 providing the drive control to two transistors in a push-pull type dc-to-dc converter. The first transformer's output is buffered by two more drive transistors—TIP35A transistors made by Texas Instruments. This switching regulator scheme uses a feedforward technique with the output voltage described by $V_{OUT} = \frac{V_{IN}}{n}$ with n = transformer ratio. When the input supply voltage to the 5560 exceeds a point (note the +14 V points in

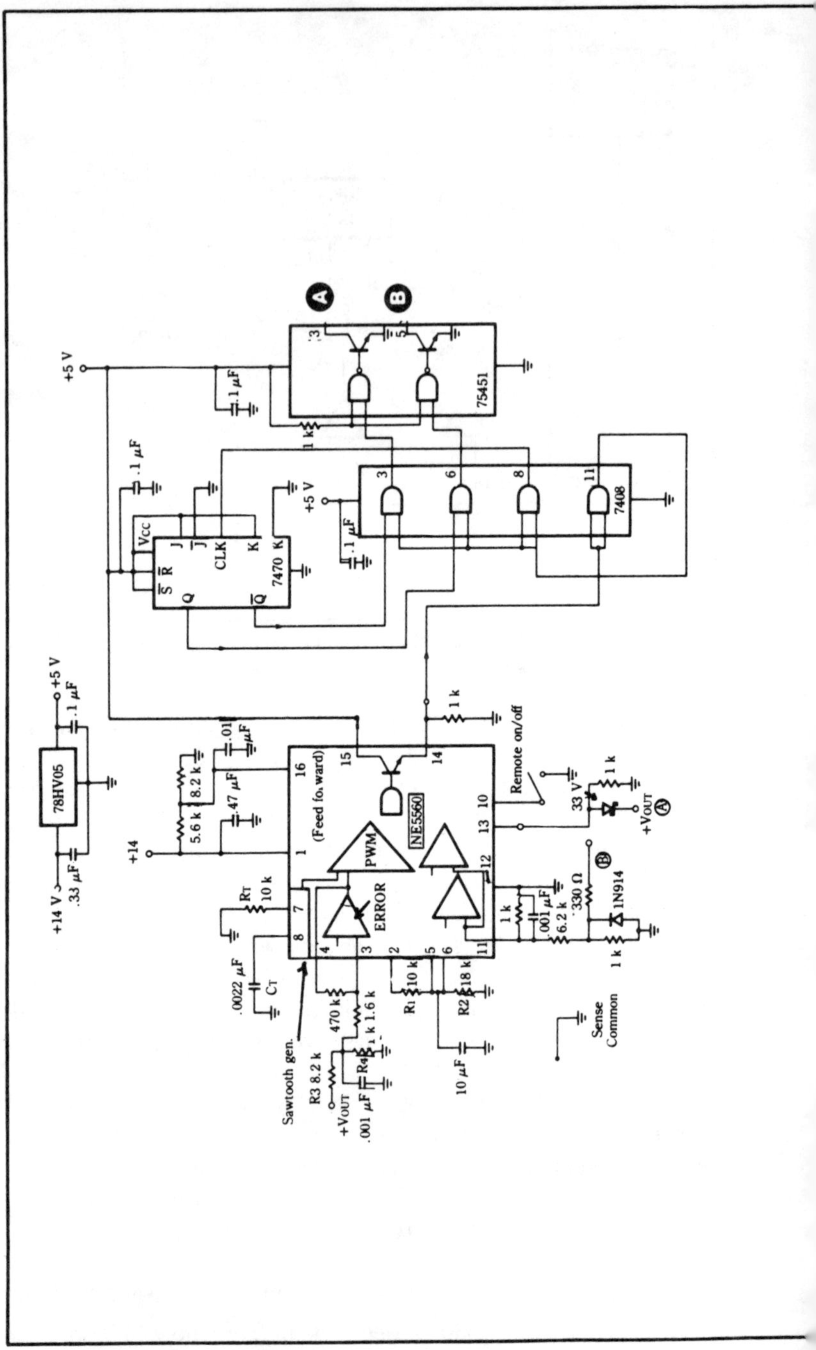

78HV05
+14 V
+5 V
.33 μF
.1 μF
7470
CLK
7408
75451
NE5560
(Feed forward)
PWM
ERROR
Sawtooth gen.
Remote on/off
Sense Common
1N914
330 Ω
+VOUT

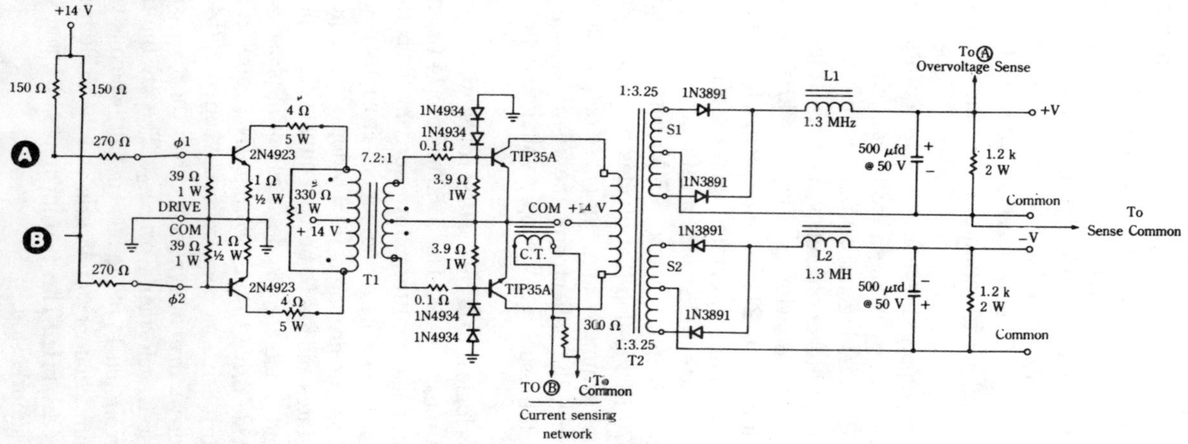

Note
Power ground and signal ground must be kept separated.

T_1, Primary = 130T (C.T.) #26
Secondary = 18T (C.T.) #22
Core = Ferroxcube 3622
3C8 material
C.T. = 50T #26 on
Ferroxcube 2616 core (3C8)
F2D bobbin

T_2, Primary = 16T (C.T.) #18 Secondaries (each) 52T (C.T.)
#22,
Core = Ferroxcube 4229 3C8 material
L_1, L_2 120T #20 on single gapped EC35 Ferroxcube
core. 3C8 material

Fig. 4-18. The NE5560 driving a push-pull converter. (From Signetics data sheets, copyright 1980 by Signetics, reprinted with permission.)

Fig. 4-18) that surpasses the zener voltage at pin 2, feedforward compensation then takes place. This means that this excess voltage on pin 16 causes an increase in the charging current for the timing capacitor on pin 8. The added voltage, however, does not change the 5560 sawtooth oscillator frequency; it only increases the upper trip point for this oscillator. Feedforward compensation is a type of preregulation for load variations reflected back to the input in changing input supply voltages. The transformer in the feedforward converter must be designed not to saturate, even under transient conditions.

The switching power supply shown in Fig. 4-18 accepts a +12 V to +18 V input range while yielding a ±30 V output with 0.1 percent regulation and achieving an 81 percent efficiency at maximum load with an even higher efficiency achieved at less than maximum load. The main PWM switches at up to 48 kHz with the power switching occurring at up to 24 kHz, one-half that frequency.

There are three TTL logic integrated circuits between the 5560 and the dc-to-dc converter. These are a 7408 quad AND gate that develops a logic high output (approx. 3.5 V) when two high inputs occur, and a 7470 JK flip-flop that divides its input square wave in half and provides symmetrical out-of-phase outputs and an open-collector buffer with the collector of the 75451 capable of receiving greater than the usual +5 V supply voltage needed by TTL.

The purpose of this logic is to drive the first transformer with a pulse-width modulated square wave. The 5560 has pin 14 (normal phase control) in the form of an emitter follower driving the bottom AND gate in the 7408 pack. This gate's output drives another AND gate above it with its output driving a JK flip-flop. This type of flip-flop will alternate states and has 50 percent duty cycle outputs. These outputs are Q and $\overline{Q}$ and are 180 degrees out-of-phase with one another; they provide drive to the two top AND gates in the 7408 pack. These in turn drive an open-collector buffer, a 75451, which provides base drive that is symmetrical to the two NPN transistors on the first transformer's primary. The JK flip-flop divides the frequency from pin 14 of the 5560 in half, which is why the 5560 has a maximum frequency of 48 kHz and the drive has a maximum frequency of 24 kHz. The two inductors, L1 and L2, provide further ripple suppression.

SG1524

The next Signetics switching regulator IC is the SG1524, also made by Silicon General Corp. The 1524 is not a complete switching power supply subsystem like the 5560 but rather only a PWM (pulse-width modulator) control circuit IC. The SG1524 comes in a 16-pin DIP and contains (refer to Fig. 4-19) a voltage reference, error amplifier, oscillator, PWM, dual alternating output switches and current-limiting and shut-down drive circuitry.

The SG1524 has an internal series-regulated 5.0 V voltage reference. This internal reference may be bypassed by tying pins 15 and 16 together and applying an external voltage up to 6.0 V there. This internal reference can also provide up to 50 mA of current.

The oscillator in the SG1524 uses an external RC timing circuit but the resistor (R_T) controls the output current from a constant-current source to timing capacitor C_T. The advantage of this is not depending on the very nonlinear charge/discharge curve of an RC pair. With a constant-current source, the curve is very linear. The capacitor should fall between 0.001 and 0.1 microfarads with the resistor between 1.8 and 100 kilohms. The period of the oscillator is approximately R_TC_T.

The SG1524 has dual alternating output switches so no JK external flip-flop would be required here. Also the two outputs, pins 12 and 13, can be connected in parallel for an effective 90 percent duty cycle; or for push-pull applications such as in Fig. 4-18, the

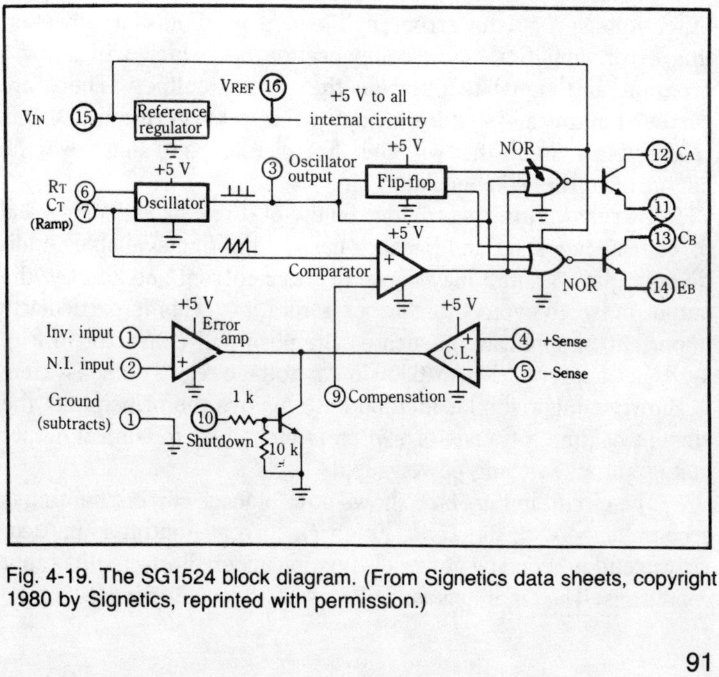

Fig. 4-19. The SG1524 block diagram. (From Signetics data sheets, copyright 1980 by Signetics, reprinted with permission.)

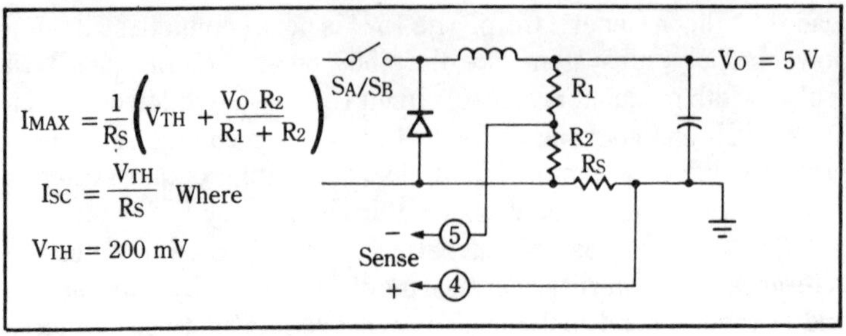

Fig. 4-20. Foldback current limiting with SG1524. (From Signetics data sheets, copyright 1980 by Signetics; reprinted with permission.)

outputs can be separated, allowing each output a 0-45 percent duty cycle. The overall frequency of the outputs is one-half that of the oscillator. If you wish to externally drive the SG1524 with your own TTL oscillator then a +3 V (TTL level) square wave clock may be applied to the oscillator output terminal with $R_T C_T$ still used and with values set slightly greater than the external input frequency. Two or more 1524 integrated circuits may be used together with one designated the master and the other the slave. The master has its $R_T C_T$ determine the oscillator frequency with the slave having its $R_T C_T$ pair set for a 10 percent longer period.

The error amplifier serves the same purpose as the 5560 and other voltage regulator error amplifiers covered thus far; however, this error amplifier has a compensation pin which will allow a programming signal to override the error amplifier. There are current-limiting and shutdown circuits connected here and all that is required is a circuit that will sink 200 microamps to shut down the output from this compensation pin.

Current limiting is provided on the SG1524 through pins 4 and 5. There is constant and foldback current limiting available. Additionally, pins 4 and 5 may sense primary current and shorten the output pulse to avoid transformer saturation. This is particularly important in applications such as the push-pull converter in Fig. 4-18. In Chapter 6, the AM6300 linear voltage regulator subsystem is shown using a similar method but employs the principle of the amount of time a transistor switch is on and off to control output voltage in a switching power supply.

The circuit in Fig. 4-20 shows how foldback current limiting is accomplished with the 1524. Foldback current limiting is particularly useful in reducing power dissipation under short-circuit output conditions. The circuit shown in Fig. 4-20 is capable of reducing the

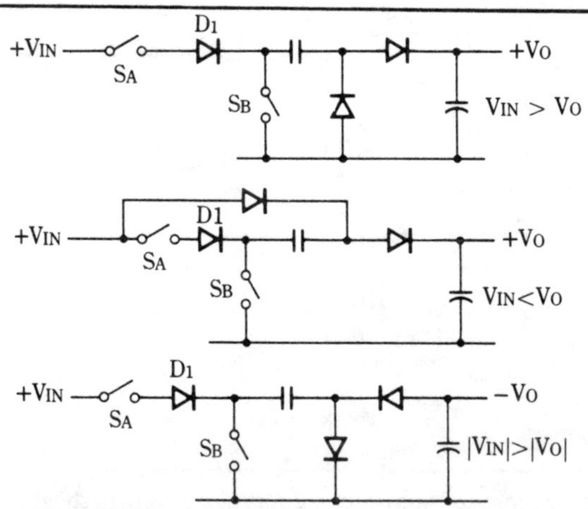

Note: Diode D1 is necessary to prevent reverse emitter-base breakdown to transistor switch SA.

Fig. 4-21. Capacitor-diode coupled voltage multipliers. (From Signetics data sheets, copyright 1980 by Signetics; reprinted with permission.)

short-circuit current to about one-third the maximum output current.

Like the 5560, the SG1524 has two identical NPN transistors with both emitters and collectors uncommitted. These transistors

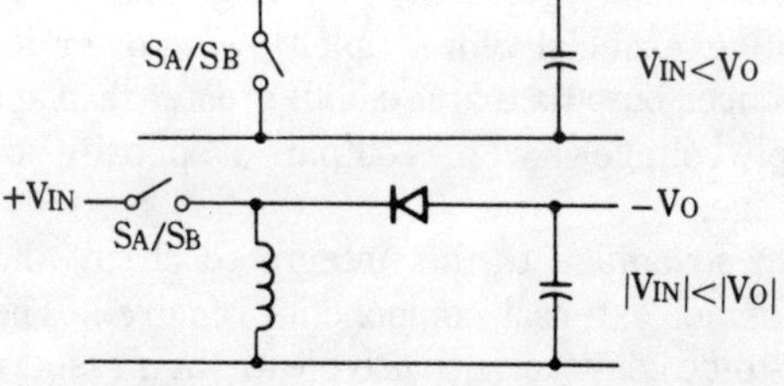

Single-ended inductor circuits where the two outputs of the SG1524 are connected in parallel.

Fig. 4-22. Inductor-capacitor single-ended circuits. (From Signetics data sheets, copyright 1980 by Signetics; reprinted with permission.)

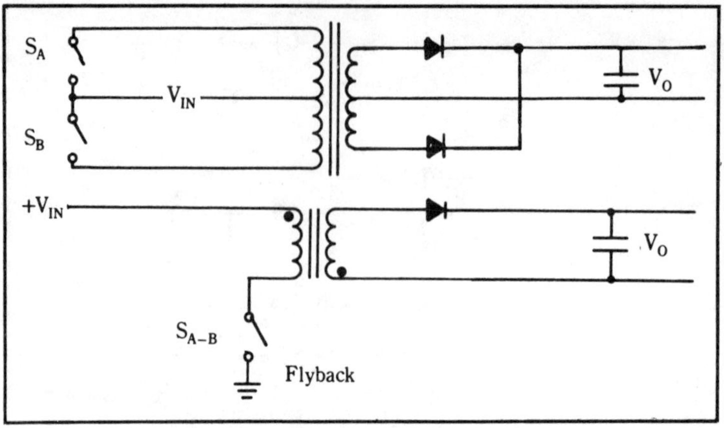

Fig. 4-23. Transformer-coupled circuits. (From Signetics data sheets, copyright 1980 by Signetics; reprinted with permission.)

are capable of sinking 100 mA for driving external power transistors. There are many possible output configurations but in general they can be grouped into the following three categories:

1. Capacitor-diode coupled voltage multipliers.
2. Inductor-capacitor single-ended circuits.
3. Transformer-coupled circuits.

Figures 4-21, 22, and 23 show these three configurations, respectively.

μA78S40

From the preceding discussion you will probably agree that the Signetics SE5560 and SG1524 are fairly sophisticated integrated circuits. These integrated circuits may be more sophisticated than many of your designs require; therefore, you might consider the Fairchild μA78S40 Universal Switching Regulator Subsystem as a viable alternative, especially for simpler lower power designs. The μA78S40 produces outputs that are both greater than and less than the input supply voltage as well as outputs of opposite polarity from the input voltage.

The main advantage of this integrated circuit, however, is reduced number of external components required. There are no traditional dc-to-dc converters to drive with their associated inductors nor are there any power transistors to drive through their bases. The μA78S40 is capable of outputs from 1.3 V to 40 V at peak noncontinuous current of 1.5 A through either the power diode and/or the power switch transistor. Note Fig. 4-24 and observe the simplicity of the design. This simplicity is not achieved at the expense of flexibility or control, because there is a spare uncommitted op amp on the IC at pins 4 through 7 which may be used to create any number of clever control and protection or voltage-level detection schemes.

Figure 4-25 illustrates the three switching modes associated with the μA78S40. These are step-down, step-up, and inverting the output voltage from that of the input voltage. All three configurations possess a switch, an inductor (L), a capacitor (C), and a diode (D1). The manner in which these three discrete passive components are placed in conjunction with the switch (the voltage regulator's output transistor) determines the mode of switching, that is, step-up, step-down, etc. Figure 4-25A demonstrates step-down switching. Here, the switch closes and the applied voltage rises toward the input voltage. The voltage $V_A - V_{OUT}$ appears across the inductor, L. This causes current in the inductor to rise, and this current flows to the load/output capacitor. If the instantaneous inductor current exceeds the load current, this excess current flows into the capacitor. Conversely, if this inductor current is less than the load current, the additional required current is drawn from the capacitor. This allows for a smoother, more continuous output.

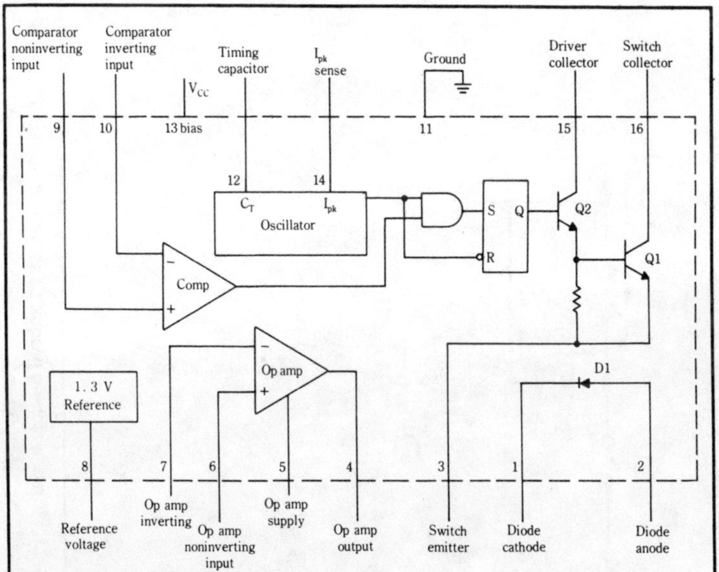

Fig. 4-24. The μ78S40 block diagram. (From the *Fairchild Voltage Regulator Handbook*, copyright 1978 by Fairchild; reprinted with permission.)

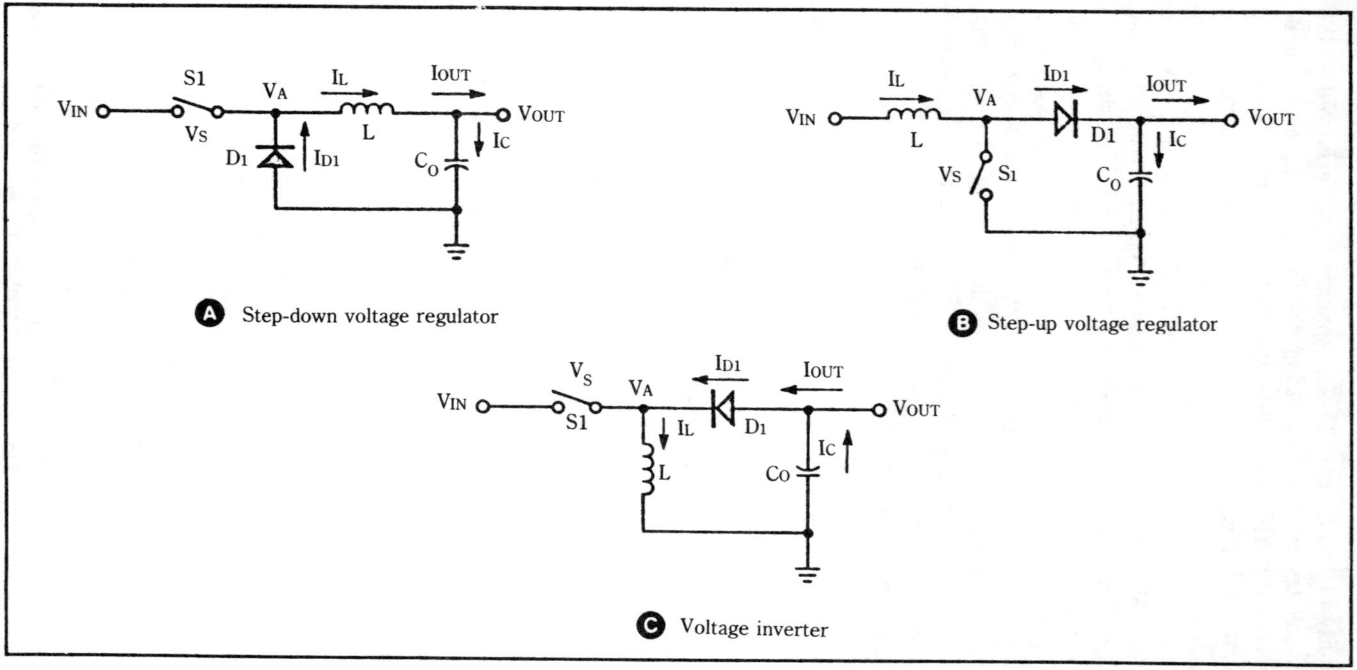

Fig. 4-25. The three switching modes possible with the μA78S40. (From the *Fairchild Voltage Regulator Handbook*, copyright 1978 by Fairchild; reprinted with permission.)

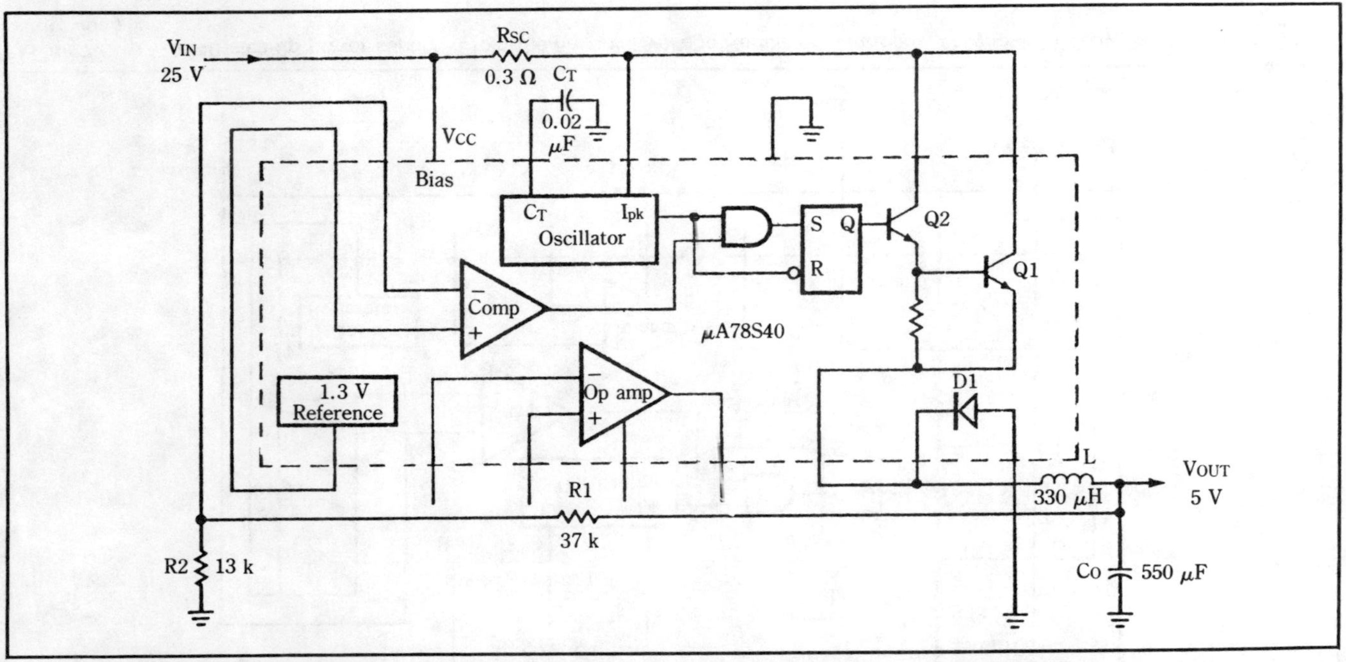

Fig. 4-26. A μA78S40 step-down power supply. (From the *Fairchild Voltage Regulator Handbook*, copyright 1978 by Fairchild, reprinted with permission.)

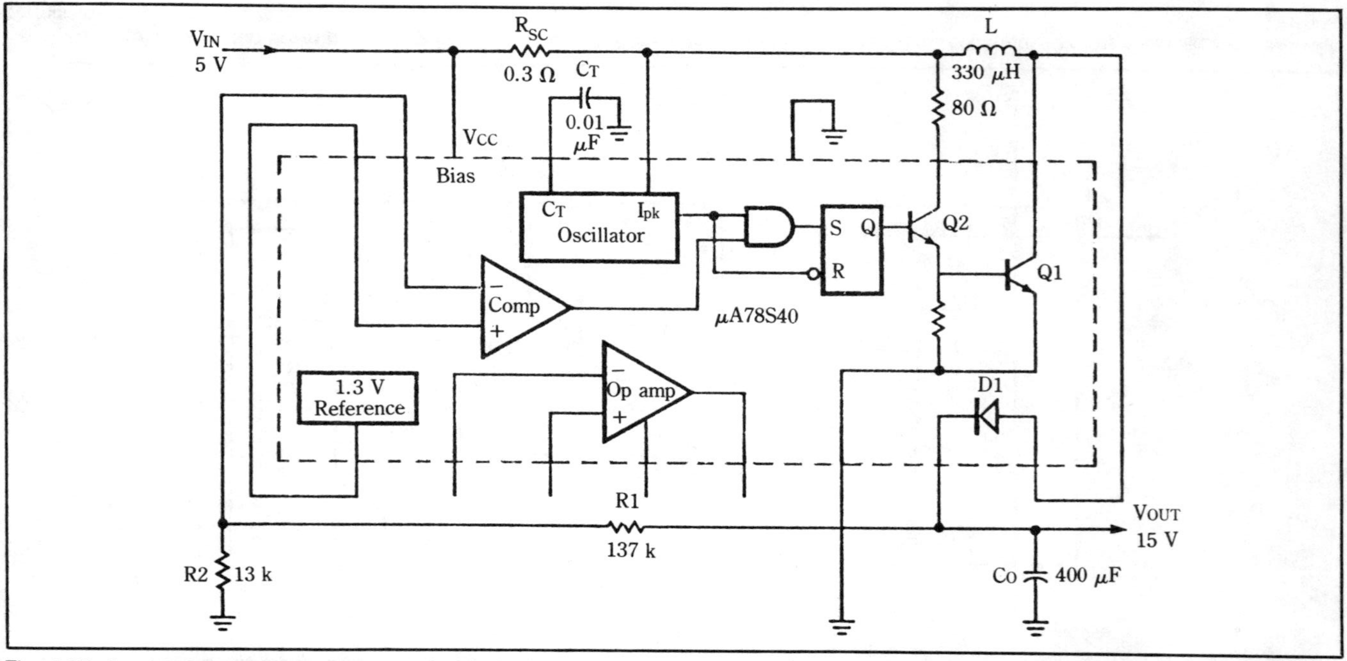

Fig. 4-27. A μA78S40 step-up power supply. (From the *Fairchild Voltage Regulator Handbook*, copyright 1978 by Fairchild; reprinted with permission.)

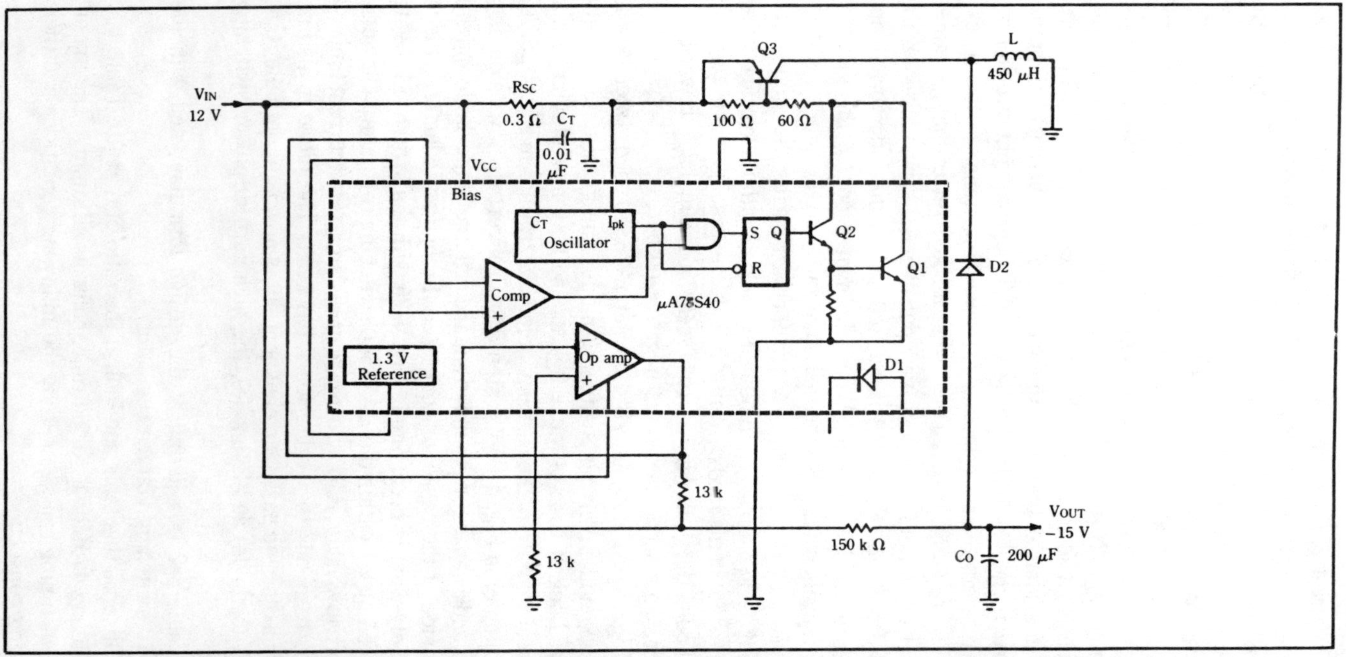

Fig. 4-28. A μA78S40 inverting power supply. (From the *Fairchild Voltage Regulator Handbook*, copyright 1978 by Fairchild; reprinted with permission.)

Figure 4-25B demonstrates the step-up switching mode. Upon closing the switch, the applied voltage drops to almost zero and a voltage of $V_{IN} - V_S$ is applied across the inductor, L. Since the applied voltage is less than the output voltage, the diode is reverse-biased and current cannot flow to the output. When the switch opens, the inductor current cannot change instantly and the applied voltage changes to the total of the output voltage plus the diode voltage. Now the diode conducts, allowing the load capacitor to charge. The inductor current now decreases at a linear rate. Timing adjustments control the average diode current, attempting to make it equal to the load current so that huge current inrushes are not experienced at the instant the switching transistor switches.

Figure 4-25C shows a switching-mode inverter that generates a negative output from a positive input. When the switch closes, the applied voltage ($V_{IN} - V_S$) is applied across the inductor, L, and causes it to rise linearly. When the switch opens, the inductor current again cannot change instantaneously so the applied voltage decreases to ($V_{OUT} - V_D$), which forward biases the diode. The current decays linearly and flows to the load with its average value equal to that of the load current. Note the reversal in the direction of the diode, thus allowing it to "steer" current back toward the source. This accounts for the negative output.

Figures 4-26, 27, and 28 show the μA78S40 used in step-down, step-up, and inverting switching-mode configurations, respectively. Look at each of these figures and note how similar, if not identical they are to the corresponding illustrations in Fig. 4-25. As an example, in Fig. 4-25A, a step-down converter shows a diode's anode and a capacitor grounded and the cathode of the diode and the other side of the capacitor connected through an inductor. This is precisely what is occurring in Fig. 4-26 with the switch, S1, being replaced by Q1 within the μA78S40 switching regulator. The addition of a low value, R_S or short-circuit resistor, is for current limiting and corresponds to a value for 1.1 A peak. The diode is also no longer a discrete passive component but is now built onto the IC at pins 1 (cathode) and 2 (anode), the high current power-switching diode.

Table 4-2 is a listing of the design formulas used with the μ78S40 switching voltage regulator.

Observe this table and look at the ratio of t_{on}/t_{off} in Table 4-2. This, along with the t_{off} or off time, controls all timing. The importance of timing can be seen by the following conditions and their resulting consequences.

●On-time too short. Inductor peak current isn't high enough, so the maximum output current is reduced.

●On-time too long, or off-time too short. Without current limiting, after several cycles the inductor current will exceed the expected peak current and continue to increase, which could lead to excessive current flow and DEVICE DESTRUCTION.

●Off-time too long. Inductor current falls to zero and stays there for some time before the switch turns on again. Consequently, average inductor current is less than half the peak current, and the available output current is reduced.

●On-time too long. Current limiting for the switch is unacceptable when the inductor current is constant at the peak current, because inductor voltage drops to zero. Therefore, switch voltage increases to $(V_{IN} - V_{OUT})$, creating a large efficiency loss and possibly DESTROYING THE SWITCH.

●Off-time too short with on-time adjusted by current limit. The inductor current does not drop to zero; therefore, average inductor current is in excess of half of the peak current, and the maximum output current exceeds half of the peak current.

Table 4-2. Design Formulas for the µA78S40. (From the *Fairchild Voltage Regulator Handbook*, copyright 1978 by Fairchild; reprinted with permission.)

Characteristic	Step Down	Step Up	Inverting
I_{pk}	$2\,I_{OUT}(max)$	$2\,I_{OUT}(max \cdot \frac{V_{OUT} + V_D - V_S}{V_{IN} - V_S}$	$2\,I_{OUT}(max) \cdot \frac{V_{IN} + \lvert V_{OUT} \rvert + V_D - V_S}{V_{IN} - V_S}$
R_{SC}*	$0.33\ V/I_{pk}$	$0.33\ V/I_{pk}$	$0.33\ V/I_{pk}$
$\frac{t_{on}}{t_{off}}$	$\frac{V_{OUT} + V_D}{V_{IN} - V_S - V_{OUT}}$	$\frac{V_{OUT} + V_D - V_{IN}}{V_{IN} - V_S}$	$\frac{V_{OUT}\rvert + V_D}{V_{IN} - V_S}$
L*	$\frac{V_{OUT} + V_D}{I_{pk}} \cdot t_{off}$	$\frac{V_{OUT} + V_D - V_{IN}}{I_{pk}} \cdot t_{off}$	$\frac{\lvert V_{OUT} \rvert + V_D}{I_{pk}} \cdot t_{off}$
t_{off}	$\frac{I_{pk} \cdot L}{V_{OUT} + V_D}$	$\frac{I_{pk} \cdot L}{V_{OUT} + V_D - V_{IN}}$	$\frac{I_{pk} \cdot L}{\lvert V_{OUT} \rvert + V_D}$
CT*(µF)	$45 \times 10^{-5}\, t_{off}(\mu S)$	$45 \times 10^{-5}\, t_{off}(\mu S)$	$45 \times 10^{-5} t_{off}\ (\mu S)$
C_O *	$\frac{I_{pk} \cdot (t_{on} + t_{off})}{8\, V_{ripple}}$	$\frac{(I_{pk} - I_{OUT})^2 \cdot t_{off}}{2\, I_{pk} \cdot V_{ripple}}$	$\frac{(I_{pk} - I_{OUT})^2 \cdot t_{off}}{2\, I_{pk} \cdot V_{ripple}}$
Efficiency	$\frac{V_{IN} - V_S + V_D}{V_{IN}} \cdot \frac{V_{OUT}}{V_{OUT} + V_D}$	$\frac{V_{IN} - V_S}{V_{IN}} \cdot \frac{V_{OUT}}{V_{OUT} + V_D - V_S}$	$\frac{V_{IN} - V_S}{V_{IN}} \cdot \frac{\lvert V_{OUT} \rvert}{\lvert V_{OUT} \rvert + V_D}$
$I_{IN(avg)}$ (Max load condition)	$\frac{I_{pk}}{2} \cdot \frac{V_{OUT} + V_D}{V_{IN} - V_S + V_D}$	$\frac{I_{pk}}{2}$	$\frac{I_{pk}}{2} \cdot \frac{\lvert V_{OUT} \rvert + V_D}{V_{IN} + \lvert V_{OUT} \rvert + V_D - V_S}$

Table 4-3. Bipolar vs MOS Transistor Tradeoffs.

Mosfet

1. FET switches are much faster—10 to 20 nsec switching times can be readily achieved.
2. No storage time effect.
3. Very low drive power required.
4. Mosfets are more expensive at present, but should be much cheaper in the future.
5. There is limited selection of Mosfet power devices.

Bipolar

1. Transition times of 50 to 200 nsec at best, with careful base drive design.
2. Storage times of 500 to 1500 nsec common. This can be reduced to 150 to 300 nssec with antisaturation circuits.
3. Base drive requirements relatively high, especially with base overdrive for fast switching.
4. At present bipolars, even fast ones, are lower in cost.
5. A very wide variety of high speed and high power devices are available.

HIGH-FREQUENCY SWITCHING MODE DESIGNS

The switching power supply designs covered thus far have been in the 20 kHz to 50 kHz range. With the introduction of the VMOS transistor acting as a "power switch" it is possible to switch amps in nanoseconds. This enables the design of switching dc-to-dc converters in the 100 kHz to 5 MHz range. Why push the switching frequency up so far? The answer is that the cost and size of the components both shrink. Low-power designs (less than 100 watts) can especially benefit from operation in the MHz range with VMOS power transistors.

Component Specification

Operation above 100 kHz requires a different look at components because of the parasitic effects which are often ignored at 20 kHz. The designer is faced with the immediate question of what type of switch to use, bipolar or MOSFET. Table 4-3 shows the tradeoffs involved. Above 100 kHz, diodes have reverse recovery

Table 4-4. Diode Operating Parameters.

Manufacturer	Type	V_R	I_F	t_{rr}
UNITRODE	UES803	150 V	70 A	50 ns
TRW	SVD450-12	450 V	12 A	70 ns
SEMTECH	3FF500	50 V	1 A	30 ns
TRW	DSR5100	1000 V	4 A	100 ns
SSDI	HSR4S	150 V	20 A	25 ns

time, t_{rr}, as their main problem. A 100 ns t_{rr} is good, but 200 ns can be tolerated if an additional power loss is permissible. Table 4-4 lists some typical diode operating parameters.

Capacitors

All capacitors have unavoidable parasitic inductance and resistance. This can result in the formation of a series-resonant circuit. Above the resonant frequency, usually 20 kHz for most larger electrolytics, the net effect is one of an inductive element. Below the 20 kHz resonant frequency, the net effect is one of a capacitive element. A capacitor also has ESR (effective series resistance), but above 20 kHz, the ESR is independent of frequency; however, as the current capability of the capacitor increases, the self-resonant frequency goes down. This is an obvious problem in higher power designs where high frequency operation is desired but the inductive reactance precludes operation at these higher frequencies. One method to overcome this is to parallel several smaller high self-resonant frequency capacitors to yield lower effective inductances.

Transformers

There appears to be no special advantage in the various transformer core shapes. But while standard magnetic wire can be used in winding high-frequency transformers, there is a definite advantage in using Litz *multifilar* wire. Before explaining this term we should first recognize that while operating at higher frequencies there is a tendency for the current to go to the outside or surface of the conductor, or wire in this case. This phenomenon is called *skin effect*. If wire is made very fine and braided, the total surface area per conductor would be much greater than the surface area of a solid strand of wire. This is why Litz wire is composed of extremely fine strands held in a braided bundle. Multifilar means that the conductor is first wound in a clockwise fashion and then again in a counterclockwise manner so that the two fields cancel one another as a result of the two equal but opposing current flows. The winding of these high frequency transformers is a critical task. Sometimes the layers of the primary and then secondary are interleaved; to achieve even better coupling the primary and secondary are wound with twisted or co-axial transmission line. Many of the principles of RF transmission line transformers can be applied here.

SUMMARY

Both low and high-frequency switching-mode power supplies

Table 4-5. Common Switching Voltage Regulator ICs. (From *Fairchild's Semiconductor Progress,* copyright 1982 by Fairchild; reprinted with permission.)

Manufacturer Part No.	Max V_{IN} (V)	Min V_{IN} (V)	Max Supply Current (mA)	Ref Volt. (V)	Ref Tol. (%)	Ref. Temp. Stability (ppm/°C)	Max Osc Freq. (kHz)	Max (Switch) Current	Switch V_{Sat}	Diode Current	Diode Forward Drop	Current Limit Sense Volt.	Features**
Fairchild μA78S40	40	2.5	5	1.245	5.2	—	100	1.5 A pk	1.3 V at 1 A	1.5 A pk	1.5 V at 1 A	330 mV	H
Fairchild μA494	42	7	10	5.0	5	300	300	250 mA	2.5 V at 250 mA	—	—	—	D,E,F,H,K
Fairchild SH1605	40	8	30	1.2	5.2	—	50	8 A	2.0 V at 5.0 A	8 A	2.0 V at 5 A	—	—
Texas Instruments TL497	15	4.5	10	1.2	5	—	—	750 mA	1 V at 500 mA	750 mA	1.75 V at 0.5 A	700 mV	H,K
Silicon General SG1524	40	8	10	5	4	100	300	100 mA	2 V at 50 mA	—	—	—	D,E,F,H,K
Silicon General SG1525	35	8	20	5.1	1	98	400	100 mA	2 V at 100 mA	—	—	—	D,E,F,K
Silicon General SG1525A	35	8	20	5.1	1	98	400	100 mA	2 V at 100 mA	—	—	—	C,D,E,F,K

Silicon General SG1526	35	8	18	5	1	100	400	100 mA	1.2 V at 100 mA	—	—	—	C,D,E,F,K
Silicon General SG1527	35	8	20	5.1	1	98	400	100 mA	2 V at 100 mA	—	—	—	D,E,F,K
Silicon General SG1527A	35	8	20	5.1	1	98	400	100 mA	2 V at 100 mA	—	—	—	C,D,E,F,K
Siemens TBA4700	30	10	—	2.5	6	100	100	70 mA	1 V at 25 mA	—	—	—	A,C,D,E, H,J
Motorola MC3420	30	10	22	7.8	5.1	300	200	50 mA	0.5 V at 40 mA	—	—	—	A,D,E,F,K
Signetics NE5560	30	NA	10	3.7	8.8	100	100	40 mA	0.5 V at 40 mA	—	—	480 mV	B,D,E,F, H,I,J,K
Ferranti ZN1066	NA*	NA*	40	2.52	2	80 (Max)	500	60 mA	0.4 V at 60 mA	—	—	—	A,D,G,K
Lambda LAS3800	40	12	13	1.65	5	80 (Typ)	150	500 mA	2 V at 500 mA	—	—	—	A,D,E,F H,K
Thomson 1001	14	6.25	18	2.5	—	80 (Typ)	50	3 A	—	—	—		C,D,E,F, H,J

Table 4-5. Common Switching Voltage Regulator ICs. (From *Fairchild's Semiconductor Progress*, copyright 1982 by Fairchild; reprinted with permission.) (Continued from page 105.)

Feature	Purpose	Advantages
A. Dynamic volt time balancing or symmetry correction.	Matches output pulse width.	Eliminates transformer unbalance, preventing transformer saturation, which can draw excessive transistor current.
B. Feedforward	Adjusts duty cycle based on input voltage rather than feedback from output.	Improved line regulation and line transient response.
C. Undervoltage lockout	Turns off control IC when supply voltage is too low.	Prevents output malfunction with input voltages less than that required for normal operation.
D. Synchronization terminal	Allows several control ICs to operate at the same switching frequency.	Reduces switching noise and simplifies input and output filtering.
E. Soft start	On initial turn on, holds outputs off, then slowly increases pulse width to normal value.	Prevents control IC from turning on hard initially in an attempt to meet output current demand.
F. Dead-time control	In a push-pull configuration, this circuit establishes a minimum off time.	Prevents output malfunction due to simultaneous pulsing of the two output push-pull transistors.
G. Dynamic dead-time control	Senses the switching state of the power transistors and automatically varies the dead time to prevent simultaneous pulsing.	Adjusts the dead time in response to changes in temperature, duty cycle, load current, etc. Static dead-time control establishes a minimum dead time which can limit maximum duty cycle especially at higher switching frequency.
H. Current-limit short-circuit protection	Removes drive to the output stage transistors.	Protects the control IC from accidental short circuit or overload.
I. Loop fault protection	Adjusts the switching duty cycle to a low or zero value when an open or short-circuit condition s present.	Without loop fault protection, the control IC may go to maximum duty cycle under open or short condition. Maximum duty cycle can be destructive in this situation.
J. Overvoltage protection	Shuts down output of control IC when output voltage exceeds predetermined value.	Prevents output overvoltage which can be fed on to the system being powered.
K. Shutdown	Allows a TTL level signal to turn off the output stage, in effect to turn off the regulator.	Microprocessors or other control logic can be interfaced with the control IC and govern its operation. External circuitry can sense fault conditions and use the shutdown terminal to protect the supply.

have been covered in this chapter as well as three integrated circuits that drive and control these power supplies. Table 4-5 is a listing of the integrated circuits that are available for driving and/or controlling these types of power supplies.

Chapter 5

60 Hz dc dc

The 723 Voltage Regulator

The 723 precision voltage regulator has been around for more than ten years and has become an industry standard building block for power supply designs. The regulator costs less than a dollar and is available at virtually any store that sells electronic components. It is capable of output voltages from 2 to 37 volts at currents up to 150 mA. There are methods of using external components to extend both this voltage and current range. The 723 is a classic regulator with an error amplifier, temperature-compensated voltage reference, series-pass transistor and current-limiting circuitry built on the IC itself. The error amplifier can be compensated for unity gain by connecting a 20 pF capacitor between the frequency-compensation terminal and the inverting input, or by connecting a 5,000 pF capacitor between the compensation terminal and the V^- terminal. When the closed-loop gain of the error amplifier is greater than unity or V_{REF}, the compensation capacitor can be decreased accordingly. Refer to the 723 data sheet in Appendix.

The 723 comes in three different packages, the round TO-100 "tin can," the plastic DIP (dual inline package), and the ceramic or military temperature range DIP. The DIP is capable of dissipating one watt at room temperature and the TO-100 metal "can" can dissipate 850 mW under similar conditions. A simplified diagram of the 723 is shown in Fig. 5-1. Note that this device contains an error amplifier, constant-current source (in the form of the double intersecting circles) and a voltage-reference zener. The output circuitry consists of a series-pass transistor whose emitter is the V_{OUT} pin. The other transistor and zener diode form a unique pair, affording

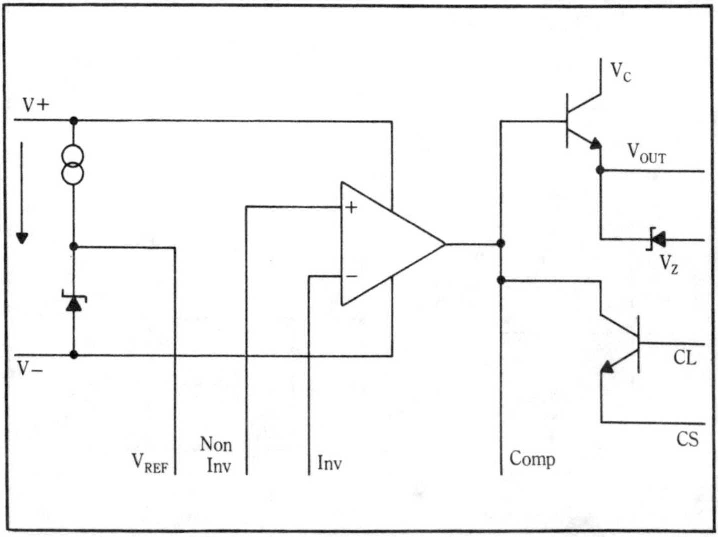

Fig. 5-1. A simplified 723 voltage regulator. (From the National Semiconductor Voltage Regulator Handbook, copyright 1982, by National Semiconductor; reprinted with permission.)

the designer considerable latitude in configuring different current-limiting and sensing schemes. The current-limiting transistor can even be used as a zener by back-biasing the emitter-base junction. The result is a 6.2 V zener with a maximum zener current of 5 mA. The back-biasing of the emitter-base junction of any transistor until it experiences zener breakdown is a perfectly legitimate design procedure; however, once a transistor has been used in the breakdown mode, it has very poor electrical noise qualities and its frequency response is also adversely affected. So if you need to make a zener diode from a transistor, it is quite acceptable as long as that same transistor is not used in a signal amplification application later.

Some of the circuits in the *IC Array Cookbook* demonstrate how voltage regulators and related circuits can be made with IC transistor arrays, some of which are used in the zener mode.

The output circuitry can be configured in three different ways, as shown in Fig. 5-2. The zener diode, D1, is a 6.2 V reference that is accessible. The base-emitter junction of transistor Q2 can also be configured as a 6.2 V zener diode, as previously mentioned. This voltage, 6.2 V, is where the zener and avalanche phenomena approximately equal one another, so this yields zeners of near zero temperature coefficients.

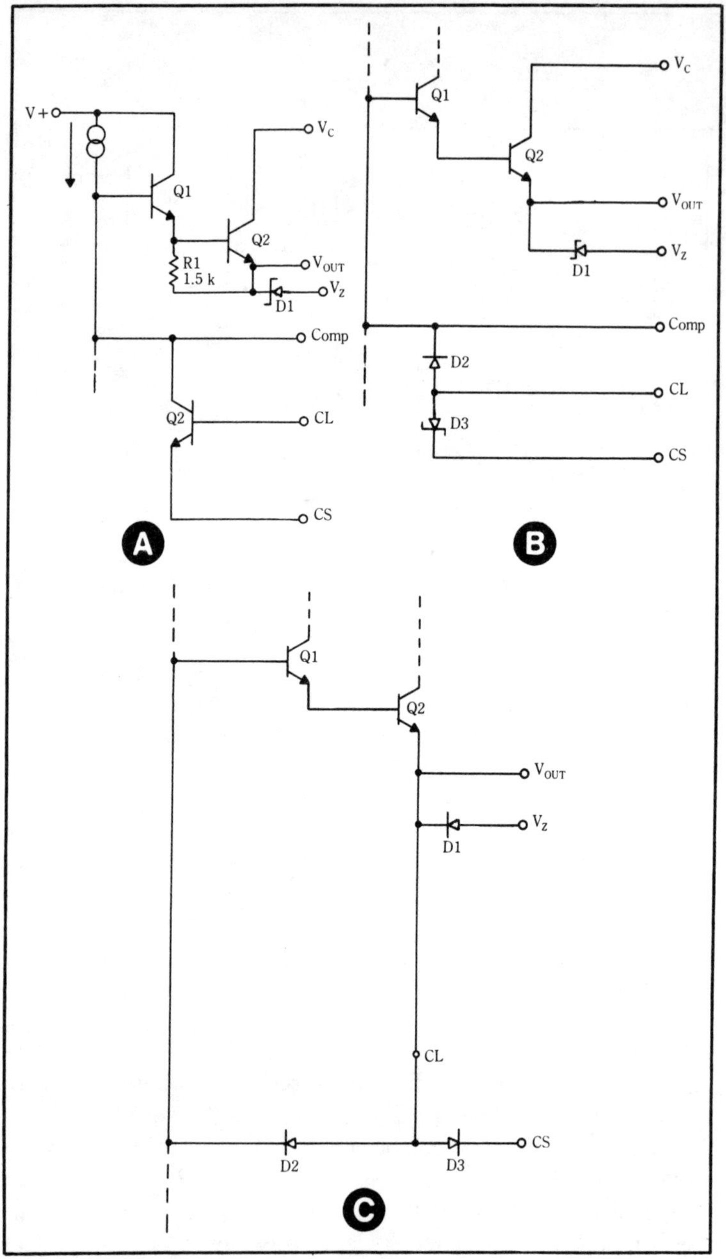

Fig. 5-2. The 723 output circuitry. (From the *National Semiconductor Voltage Regulator Handbook*, copyright 1982 by National Semiconductor; reprinted with permission.)

A BASIC ADJUSTABLE POWER-SUPPLY DESIGN

Because of the ever-increasing complexity of integrated circuits, there is a tendency to view designs from a black box or functional block approach. This is particularly true of operational amplifiers. It is perfectly acceptable to do this to quickly come up with a circuit that has the proper output voltage and operational parameters. These may be obtained, in the form of design equations, from the illustrations that follow. If your curiosity goes beyond this, you can do two things that will greatly improve your understanding of voltage regulators and specifically the 723. First, draw the 723 in the functional symbol manner as demonstrated in Fig. 5-1, but be sure to number the pins. Next, take the application circuit you intended to use, such as in Fig. 5-3 and redraw it as has been done in Fig. 5-4. This circuit is now easier to understand by using this method of "expanding and redrawing".

Figure 5-4 contains the schematic for a variable-output voltage, high-current power supply. Transistor Q1 is the series-pass transistor through which all load current must pass. Looking at the internal circuitry of the 723, note that the constant-current source biases the 6.2 V zener diode. By maintaining a constant current through the temperature-compensated zener, its 6.2 V output volt-

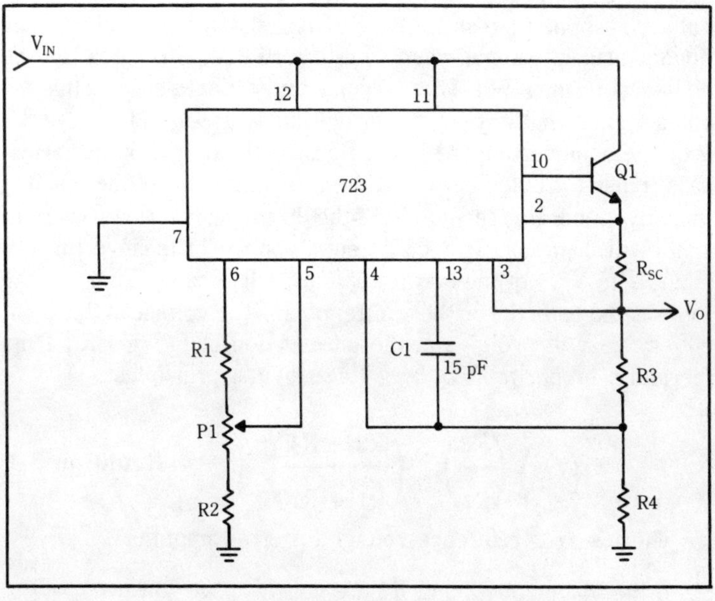

Fig. 5-3. An adjustable current-boosted 723 power supply.

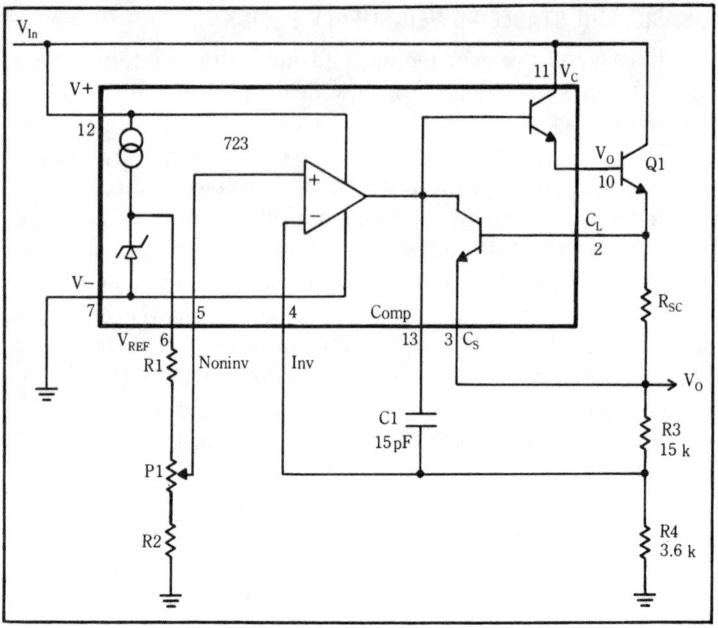

Fig. 5-4. A 723 regulator redrawn for clarity.

age will remain much more constant. This constant 6.2 V zener voltage is applied across pot P1, a 5 k pot, and by the simple voltage divider principle some value from 0 to 6.2 V will appear on the noninverting input (+) of the error amplifier. Let us assume that the pot is set at its halfway point. This will apply one-half of 6.2 V or 3.1 V on the noninverting (+) input. Because the output of the series-pass transistor is fed back into the inverting input of the op amp, (negative feedback) the output of the op amp and series-pass transistor will change until a 3.1 V potential appears on the inverting (−) input. The V_{OUT} of the error amplifier will be approximately the same as the potential on the emitter of Q1. The voltage at this point is very near the voltage at the intersection of $R_{SÇ}$ and R_3. It is, therefore, proper to state by voltage divider principles:

$$V_{OUT} = \left(V_{REF}\right)\left(\frac{R2}{R4}\right)\left(\frac{R3 + R4}{R1 + R2}\right) \qquad \textbf{Equation 5-1}$$

where V_{REF} = reference voltage for error amplifier

Since the values of R3 and R4 are constant, we can simplify the above equation by rearrangement into:

$$V_{OUT} = V_{REF} \left(\frac{R2}{R1 + R2}\right) \underbrace{\left(\frac{R3 + R4}{R4}\right)}$$

$$\left(\frac{15\text{ k} + 3.6\text{ k}}{3.6\text{ k}}\right) = 5.16$$

$$V_{OUT} = \underbrace{V_{REF}}_{6.2\text{ V}} \left(\frac{R2}{R1 + R2}\right) (5.16)$$ **Equation 5-1A**

If V_{REF} is 6.2 V and the pot is set at its half-way point then

$$V_{OUT} = (6.2)\left(\frac{1}{2}\right)(5.16) = 16 \text{ volts}$$

Also, once the equation of the output voltage has been simplified it is obvious that the pot controls the ratio of R2/(R1 + R2) and in turn the voltage on the noninverting (+) pin which the inverting pin (−) must match in voltage.

SERIES PASS TRANSISTORS

Examining the output circuitry, the output transistor on the 723 IC and transistor Q1, the external series-pass transistor, form a Darlington pair and their common collectors go to $+V_{IN}$. This transistor configuration can be visualized as essentially a single transistor with greatly increased current gain capabilities. In fact, the current gain of a Darlington pair is very nearly the product of the current gains (beta) of the first and second transistors combined. Transistor Q1 is specified with a greater current-carrying capability than the 150 mA capability of the output transistor on the 723 regulator IC. Power transistors like this often only have gains (betas) on the order of 10 to 15.

Developments in the power transistor industry may cause silicon bipolar transistors to be replaced with VMOS transistors. A list of recent references on this subject is provided at the end of this book. In addition, *Design of VMOS Circuits with Experiments*, does an excellent job of treating this subject in great detail.

The short-circuit sensing resistor, R_{SC}, has all of the load current flowing through it. This is intentional because it has to sense the full load current in order to determine when it starts to approach dangerous limits for the components and, therefore, when the 723 needs to go into current limiting. To better understand current limiting, refer to Fig. 5-2A. Note that the base and emitter of the current-limiting transistor come to the outside world through pins CL and CS, respectively. This stands for current limiting and current sensing. If R_{SC}, which is normally placed between the emitter and base of this transistor, is short-circuited, or these two pins (CS and CL) are tied together, the 723 will output 150 mA of current; however, if some lower current is desired, the following equation is used:

$$I_{LIMIT} = \frac{V_{SENSE}}{R_{SC}} \qquad \textbf{Equation 5-2}$$

where V_{SENSE} = the voltage between pins CL and CS (Fig. 5-5).
R_{SC} = short-circuit sensing resistor

In Fig. 5-5 note that at a 50° C junction temperature, a 120 mA output would require a 0.6 V current-limit sense voltage. By applying the current-limiting equation, a 5 ohm resistor would be required for R_{SC} for a 120 mA output current capability.

Now let us consider the case of the external series-pass transistor, Q1. If this transistor has a beta of 15, this quantity is multiplied by the 120 mA established by the 5 ohm current-limit short-circuit resistor. This would yield an output current capability of (120 mA) (15) = 1.8 A.

When the current-limiting transistor has a sufficient base-to-emitter potential, it turns on. This allows it to conduct and when

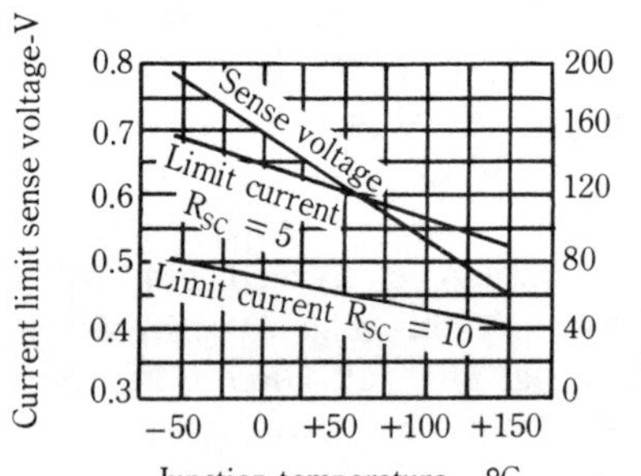

Fig. 5-5. Voltage sense graph. (From *National Semiconductor Voltage Regulator Handbook*, copyright 1982 by National Semiconductor; reprinted with permission.)

doing so pulls the output of the op amp low which cuts off the output transistor. This in turn shuts off the series-pass transistor or reduces it to a point where current-limiting goes into effect. We have been discussing regular current limiting, but foldback current-limiting will also be covered in another section of this chapter. This is a superior alternative to regular current limiting.

When using series-pass transistors, one V_{BE} drop per transistor must be added to the 3 V nominal minimum $V_{IN} - V_{OUT}$ voltage differential required by the 723 to maintain regulation. Also, a resistor may be placed between the base and the emitter of the series-pass transistor, if desired. This gives I_{CBO} or leakage current compensation.

DESIGNING A SHUNT REGULATOR

Figure 5-4 shows a series regulator application in which the full current delivered to the load passes through the series-pass transistor (the load is in series with this current path). In a shunt regulator (Fig. 5-6) all of the load current does not pass through the regulator's series-pass transistor, but passes through the collector resistor and is then divided between the load and shunt transistor, with the majority of the current passing onto the load. A positive shunt regulator with a variable output voltage is shown in Fig. 5-6. Care should be exercised to ensure that R4 is capable of handling the high power that will develop in this mode of operation. The formula for power is $P = I^2R$, and the current is the maximum load current. P is expressed in watts and a good rule-of-thumb for safety and added reliability is to double the wattage of the resistor from the calculated power value.

LOW-POWER 723 DESIGNS

In simple, low-power applications, the use of the 723 voltage regulator alone along with three or four resistors to determine output voltage and current-limiting levels is sufficient. Basically, the 723 can be configured for either low-voltage (2 V to 7 V) or high-voltage (7 V to 37 V) applications. Figure 5-7 shows the low-voltage application, and Fig. 5-8 shows the high-voltage application. Resistor R3 in both applications is not necessary if you can tolerate error amplifier drifts with temperature that will cause approximately 0.1 V excursions on the output. If a more steady output with respect to temperature change is required, then add R3. Its value should be equal to the parallel combination of R1 and R2.

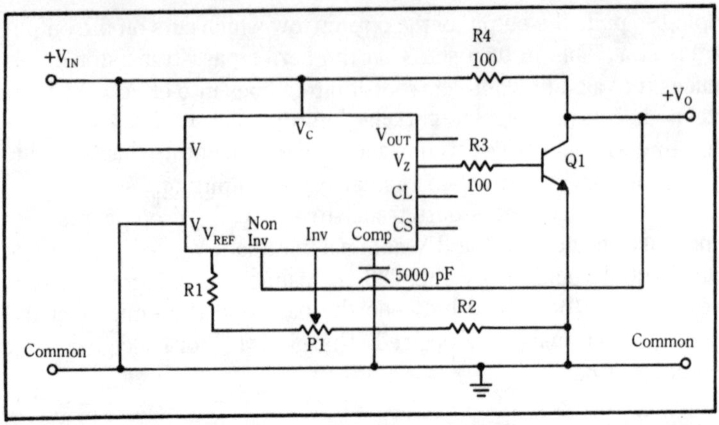

Fig. 5-6. Using the 723 as a shunt voltage regulator. (From the *National Semiconductor Voltage Regulator Handbook,* copyright 1982 by National Semiconductor; reprinted with permission.)

There are instances in which you may want to vary the 723 over a range wider than the two ranges represented, (2 V to 7 V and 7 V to 37 V), Fig. 5-9 shows how to accomplish this. The normal 7 V reference to the 723 voltage regulator's error amplifier is divided down to 2 V. The output voltage equation is now $V_{OUT} = V_{REF}$ $(1 + R_A/R_B)$ with V_{REF} now being 2 V. Current limiting is described by the equation:

$$I_{SC} = 0.6/R_{SC} \qquad \textbf{Equation 5-3}$$

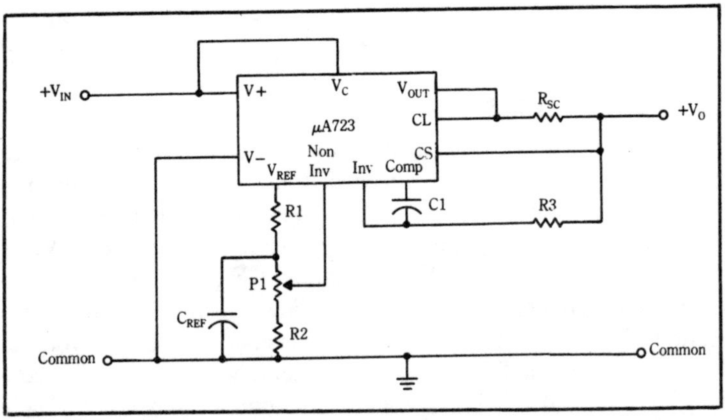

Fig. 5-7. A low-voltage (2-7 V) 723-based voltage regulator. (From *National Semiconductor Voltage Regulator Handbook*, copyright 1982 by National Semiconductor; reprinted with permission.)

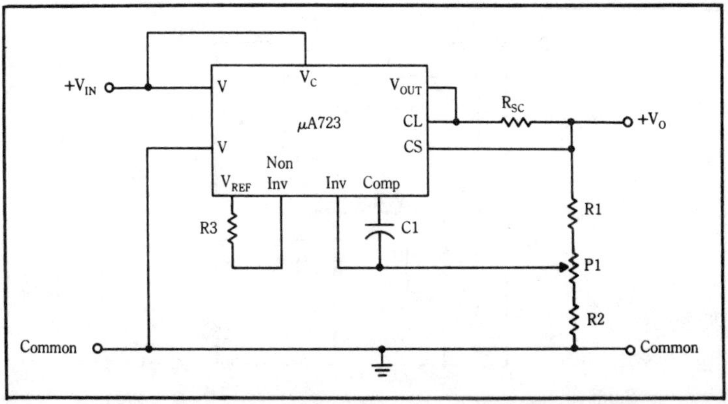

Fig. 5-8. Using the 723 for a (7-35 V) voltage regulator. (From *National Semiconductor Voltage Regulator Handbook*, copyright 1982 by National Semiconductor; reprinted with permission.)

where R_{SC} = short-circuit resistor

Pot R_A should be a ten-turn pot because the 35 V range of the regulator would be hard to adjust with a single turn pot.

723 CURRENT REGULATOR DESIGNS

A 723 voltage regulator can also be used as a current regulator. Figure 5-10 demonstrates how the regulator forces a voltage to

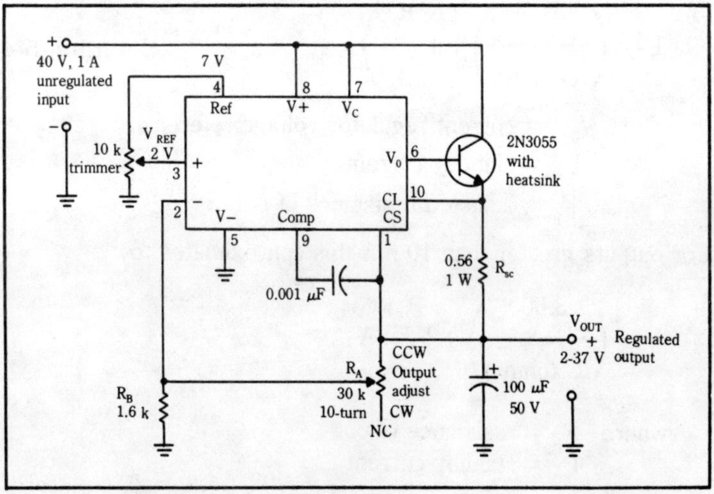

Fig. 5-9. A wide-range, adjustable, power supply. (From *EDN*, copyright 1978 by Cahners Publishing; reprinted with permission.)

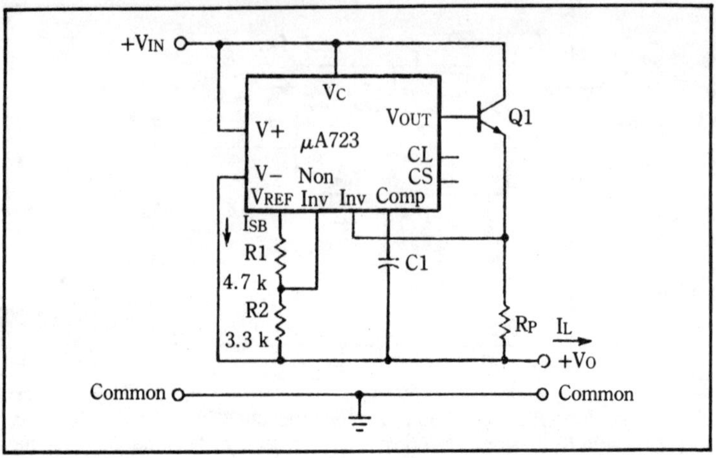

Fig. 5-10. A 723 voltage regulator used as a current regulator. (Courtesy National Semiconductor.)

appear across R_P to equal the existing voltage across R2. The resulting current through these resistors is summed with the standby current of the 723 (I_{SB}). This causes a slight inaccuracy when the current regulator is operated at levels below 10 mA; however, at elevated currents of 50 mA or more this standby current becomes more and more insignificant. The exact equation for the load current is:

$$I_L = \left(\frac{V_{REF}}{R1 + R2}\right)\left(1 + \frac{R2}{R_P}\right) + I_{SB} \qquad \textbf{Equation 5-4}$$

where V_{REF} = current regulator voltage reference
I_{SB} = standby current
R_P = adjusted resistance of pot

For outputs greater than 10 mA this approximates to:

$$I_L \simeq \left(\frac{3{,}000}{R_p \text{ (ohms)}}\right) + I_{SB} \text{ mA}$$

where R_P = resistance of pot
I_{SB} = standby current

This approximate term with 3,000 in it is valid for the values of R1 and R2 shown in Fig. 5-10. If the current regulation of Fig. 5-10 is inadequate and the limitation of a V^+ V^- differential of 40 V is also unacceptable, then the circuit of Fig. 5-11 can be used. This precision, floating current source is capable of 0.05 percent regulation. In an application like this, a floating 20 V supply (typically a half-wave rectified output from a separate transformer winding of the main supply) is used to power the regulator. This has the benefit of the standby and reference currents not adding to the programmed output current. The load current is given by:

$$I_L = \left(\frac{R2}{R1 + R2}\right)\left(\frac{V_{REF}}{R_p}\right) \qquad \textbf{Equation 5-5}$$

where V_{REF} = voltage reference of pot

R_p = resistance of pot

If V_{R2} is made to equal 3.0 V by adjusting P1, the relationship can be simplified to:

$$I_L = \left(\frac{3{,}000}{R_p \text{ (ohms)}}\right) \text{mA}$$

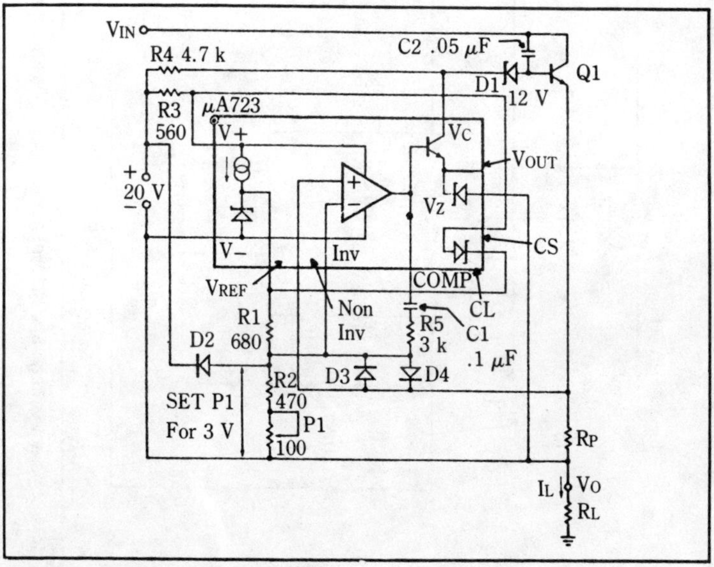

Fig. 5-11. A precision floating current source using a 723 current regulator. (courtesy of National Semiconductor.)

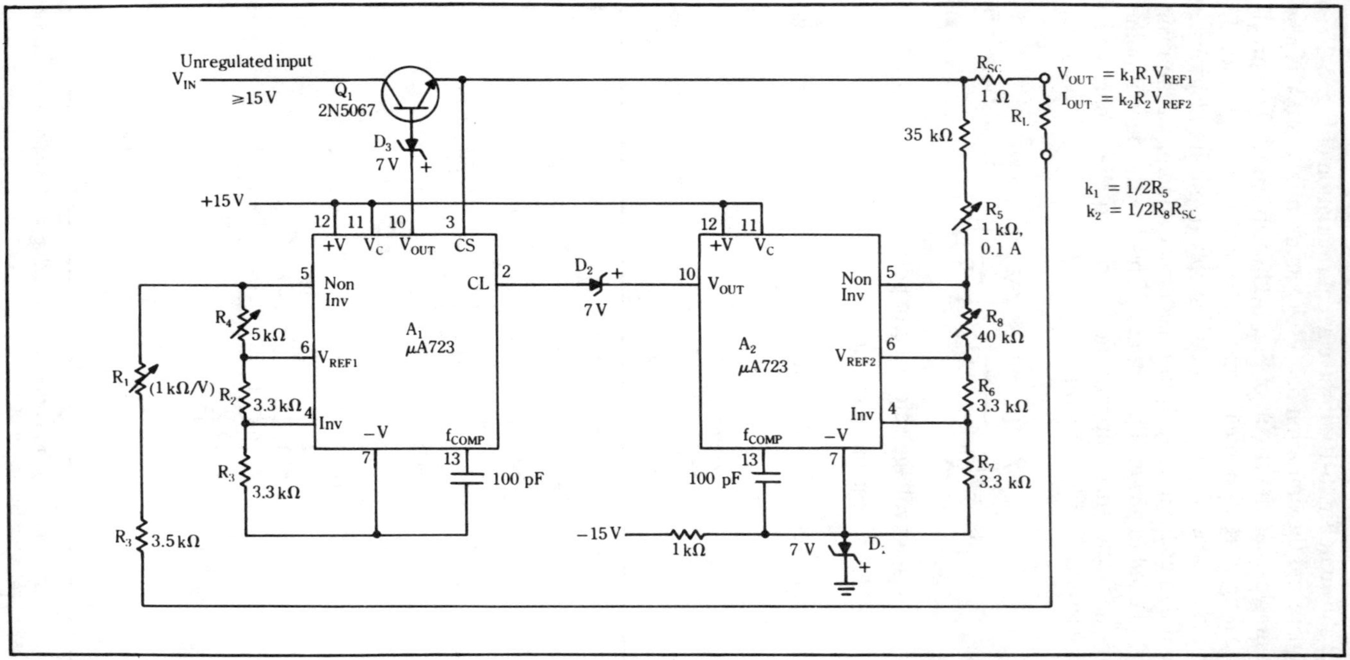

Fig. 5-12. A constant-voltage and current power supply design. (From *Electronics*, July 20, 1978, copyright 1978 by McGraw Hill; reprinted with permission.)

Figure 5-12 shows an application of two 723 regulators being cascaded to provide both a constant voltage and a constant current. The voltage may be any voltage desired over the 0 to 15 volt range with a load current that is selectable over a 0 A to 3 A range. The output voltage is controlled by 723 regulator A1 and its error-voltage amplifier (pins 4, 5, and 6). The second 723, A2, detects any change in load current and generates a correcting voltage at the V_{OUT} terminal (pin 10). The internal current-limiting transistor of A2 controls the amount of drive current supplied to series-pass transistor, Q1. The output voltage is adjusted by R1. The output current is adjusted by R5 and changes by 100 mA for every 1 kΩ change in R5, assuming that R6 = R7.

FOLDBACK VS REGULAR CURRENT LIMITING

Foldback current limiting occurs when the output current limit is reached. However, instead of the current being limited to a constant value, the current is reduced in magnitude or is "folded back." Refer to Fig. 5-13, which contains an illustration of the series-pass transistors used in this current-limiting mode plus a chart that demonstrates this phenomenon.

On the positive side, foldback current limiting allows more current drain from the regulator's output under short-circuit conditions without overstressing the series-pass elements. On the negative side, the current-sensing resistor experiences excessive power loss. As an example, if the output voltage is 5 V and the output current is 6 A and R_S = 0.4 ohms, the power loss in R_S is 12.4 watts. It is obvious that this is a major contributor to inefficiency. Figure 5-13A shows a conventional constant-current source circuit configuration with the series-pass transistor connected to points A and B. Foldback current limiting occurs when the circuit in Fig. 5-13B has its corresponding A and B points connected into the circuit in Fig. 5-13A. By adding two resistors to the normal foldback current limiting configuration shown in Fig. 5-13B, the circuit in Fig. 5-13C is developed. This circuit dramatically reduces the value of the R_S resistor while retaining the protection that is inherent in foldback current limiting. The current-limiting point may be approximated by the following equation if R_S is much smaller than R1, R2, and R3:

$$I_{OL} = [R3/R2(V_O + V_{EB})]/R_S$$ **Equation 5-6**

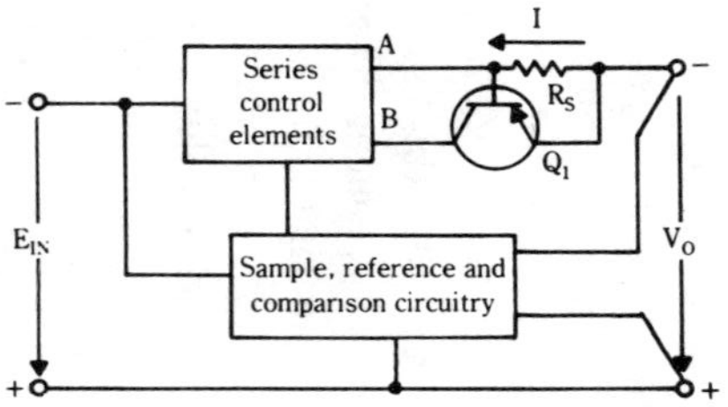

A Constant current

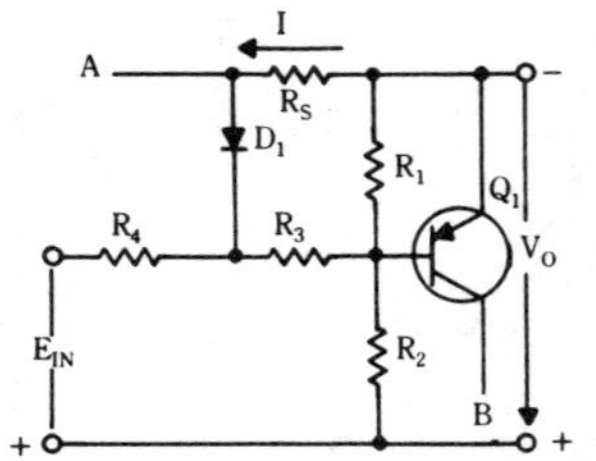

C Improved current foldback

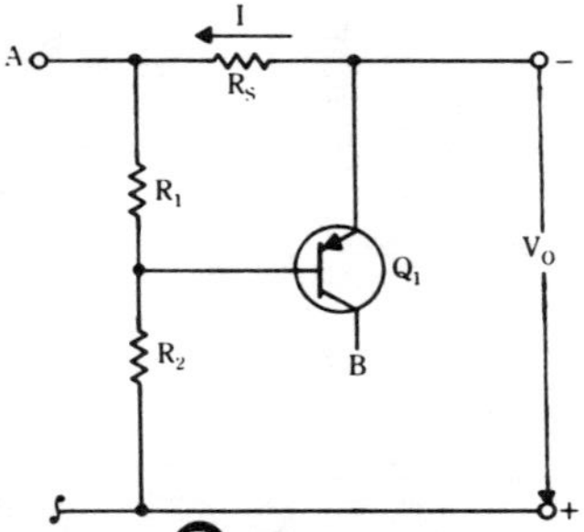

B Current foldback

Fig	(Overload (I_{OL}))	Short circuit (I_{SC})
(A)	$V_{EB\gamma}/R_S$	$V_{EB(ON)}/R_S$
(B)	$\frac{1}{R_S}\left[V_{EB\gamma} + \frac{R_1}{R_2}(V_O + V_{EB\gamma})\right]$	$\left[1 + \frac{R_1}{R_2}\right]\left[\frac{V_{EB(ON)}}{R_S}\right]$
(C)	$\frac{1}{R_S}\left[\frac{R_3}{R_2}(V_O + V_{EB\lambda})\right]$ $\frac{R_3}{R_1} = \frac{V_{D_1}}{V_{EB\lambda}} - 1$	$\frac{R_3 V_{EB(ON)}}{R_S R_2}$ $\frac{R_3}{R_1} = \frac{V_{D_1}}{V_{EB(ON)}} - 1$

D

Fig. 5-13. An alternate current foldback limiting circuit. (From *EDN*, November 5, 1975; reprinted with permission.)

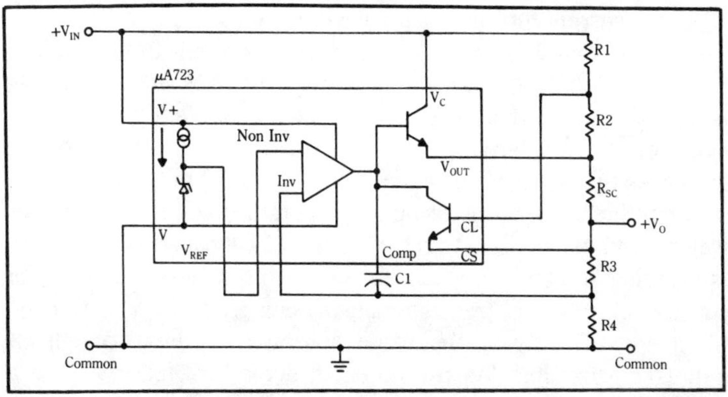

Fig. 5-14. Positive 723 foldback current limiting.

where V_{EB} = the conduction point of transistor Q1 and is approximately 0.6 V.

R_S = series resistor

Foldback current limiting is accomplished with the 723 in both a positive and negative direction in Figs. 5-14 and 5-15. This technique introduces positive feedback by increased current flow through R1 and R2 under short-circuit conditions. This forward biases the base-to-emitter junction of the current-sensing transistor. The percentage of current foldback depends on the proportion of the voltage drops of R2 and R_{SC}-to-base current drive of the current-limiting transistor. When the direct short is removed or

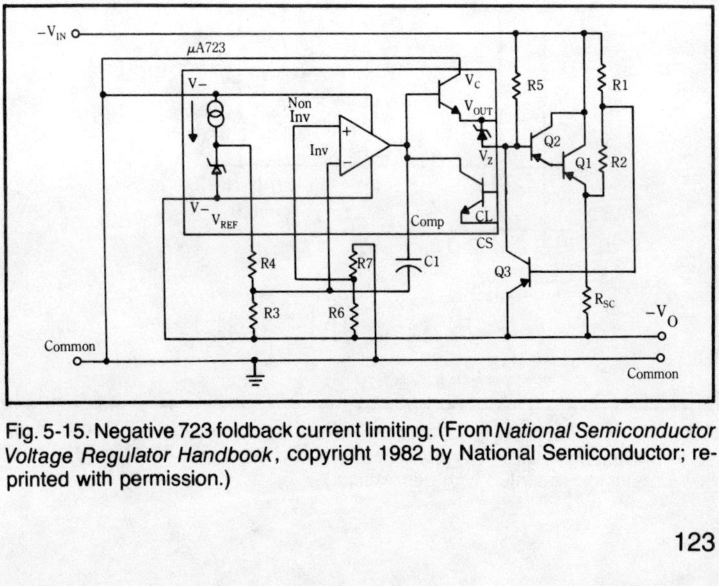

Fig. 5-15. Negative 723 foldback current limiting. (From *National Semiconductor Voltage Regulator Handbook*, copyright 1982 by National Semiconductor; reprinted with permission.)

when the current returns to a level within the safe operating region, the power supply can supply the "design" current to the load.

A problem with this configuration for foldback current limiting is latch-up. This occurs when 100 percent positive feedback takes place, or stated differently, when some change in the input voltage increases the voltage drop across R2 past the 10 mV "window" or upper tolerance span of acceptable operating voltage. A simpler way of visualizing this is to refer to Fig. 5-14 and assume that there is less than a 10 mV drop across R2. In this case, the base of the current-limiting (CL) transistor and the emitter of the V_{OUT} transistor are essentially at the same potential so they are almost "shorted" together. As the potential across R2 increases, as a result of increased input voltage, above the 10 mV limit, these two transistors latch up because of the base drive on the current-limiting (CL) transistor. Once the latch-up occurs, it can only be removed by breaking the positive feedback path by some means such as a manual reset switch, allowing the series-pass transistor to again be driven in a normal fashion. This problem can be solved by the addition of an external transistor as shown in Fig. 5-16. This transistor acts as a

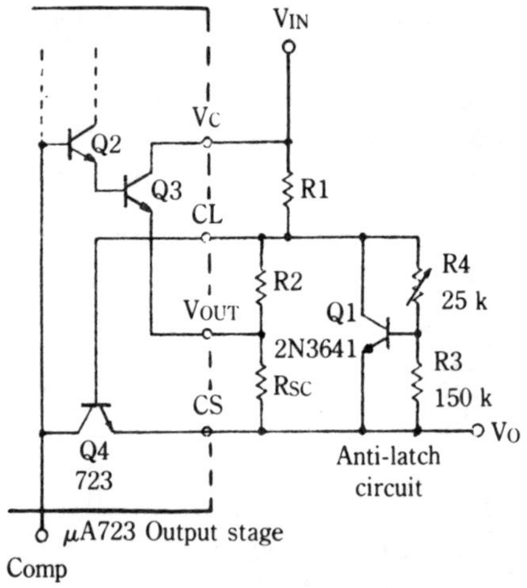

$$I_{SC} = \left[\frac{V_{SENSE}}{R_{SC}}\left(\frac{R_1 + R_2}{R_1}\right)\right] - \left[\frac{V_{IN}}{R_{SC}}\left(\frac{R_2}{R_1}\right)\right]$$

R_4 can be empirically set to a point just above the minimum output current under short circuit load and high ac line operation.

Fig. 5-16. Avoiding latch-up in foldback current limiting designs. (From *National Semiconductor Voltage Regulator Handbook*, copyright 1982 by National Semiconductor; reprinted with permission.)

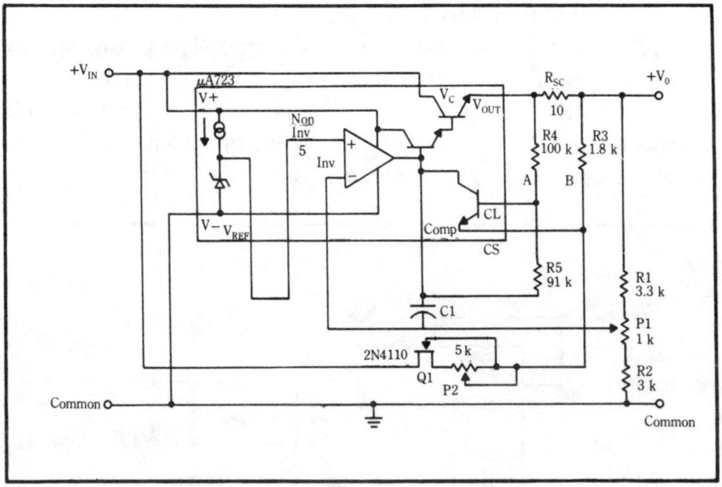

Fig. 5-17. Current limiting with temperature compensation. (From *National Semiconductor Voltage Regulator Handbook*, copyright 1982 by National Semiconductor; reprinted with permission.)

voltage regulator itself but it regulates the voltage from the base-to-emitter of the internal 723 voltage regulator's current-limiting transistor, Q4 in Fig. 5-16. Since the latching is caused by this transistor saturating, regulation of its V_{BE} is accomplished by bypassing its base current, above a certain value, set by the voltage divider R3/R4 which is applied to the base of Q1. Depending on the value of V_{IN}, R4 may be adjusted until it reaches a point that will give you the desired foldback current limiting characteristics you desire without fear of latch-up. The values of R1, R2, and R_{SC}, as well as the desired current to be limited, are provided in the equation within Fig. 5-16.

Figure 5-13 showed how to improve efficiency in foldback current limiting regulation circuits by an alternate scheme to the usual current-sense resistor. The circuit in Fig. 5-17 illustrates another circuit that dissipates very little power under short-circuit conditions. This circuit exploits the internal temperature tracking of the IC itself. More specifically, since the series-pass transistor and current-limiting transistor are physically next to one another on the IC chip or substrate and were manufactured at the same time, their base-to-emitter temperature coefficients are very nearly identical.

In Fig. 5-17, the current-limiting transistor is connected so as to cancel the temperature coefficients of the series-pass and current-limiting transistors. A bridge circuit is used (R4 and R5), which reduces the voltage drop to a 1 V_{BE} drop. The voltage across R4 is balanced by an equal and opposite voltage developed across R3 by the FET current source. The current-limiting transistor internal to the 723 is connected across the voltages across R3, R4, and R_{SC}. At room temperature, the FET current source is adjusted

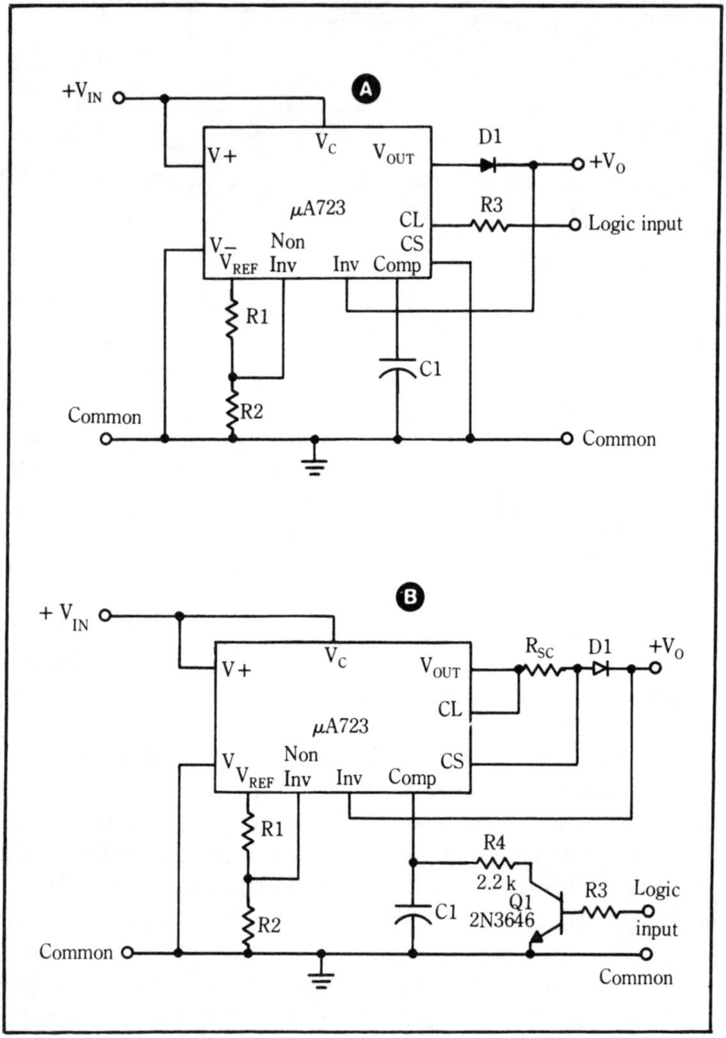

Fig. 5-18. Electronic shutdown of a 723 positive regulator. (From *National Semiconductor Voltage Regulator Handbook*, copyright 1982 by National Semiconductor; reprinted with permission.)

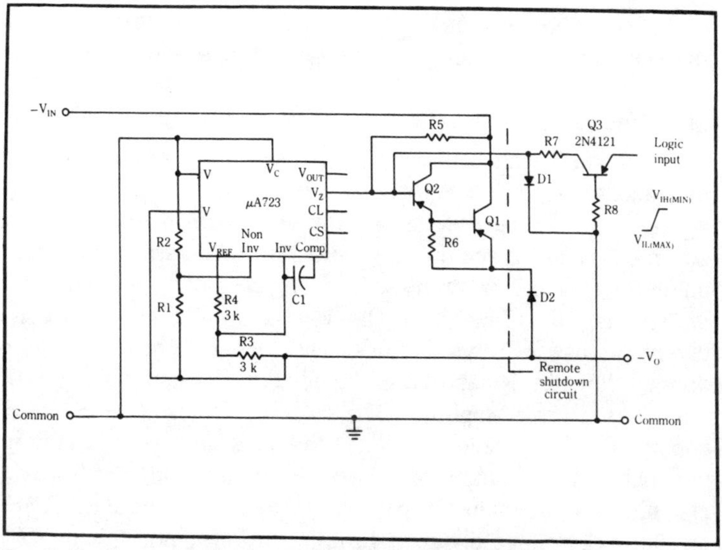

Fig. 5-19. Electronic shutdown of a 723 negative regulator. (From *National Semiconductor Voltage Regulator Handbook*, copyright 1982 by National Semiconductor; reprinted with permission.)

by pot P1 so that there is no voltage across the Cl and CS terminals of the 723. The threshold of the current-limiting transistor tracks R4 with temperature, resulting in a temperature-stabilized current-limiting range.

REMOTE REGULATOR SHUTDOWN

In Chapter 3 we investigated electronic or remote shutdown of both three-terminal regulators and the LM325/326-dual tracking voltage regulator. The 723 can also be remotely or electronically shut down by grounding the compensation pin. When grounded, the drive current to the output stage is shut off. The only requirement for the circuitry grounding the compensation pin is that it must be a positive voltage source and must be capable of sinking 100 μA of current. If direct logic drive through a resistor is used, as shown in Fig. 5-18A, R3 should be 3 k for a TTL 5 V system or 10 k for a 10 V CMOS driver. Adding diode D1 in Fig. 5-18B protects the output stage from excessive reverse base-to-emitter voltage spikes during shutdown. Remote shutdown for a 723 used as a negative voltage regulator requires some added circuitry. Note in Fig. 5-19 the added circuitry to the right of the dotted line. A logic O causes Q3 to remain off and inhibits the shutdown circuitry. A logic 1 turns Q3 on, which drives the Darlington pair of Q1 and Q2 off. The value of R8 should be great enough to turn Q3 on into saturation but limit the drive to Q3 to maintain a safe level of operation.

MONITORING VOLTAGE LEVELS

Chapter 6 deals with monitoring the power supply's health and introduces two power supply monitoring systems each contained on a single IC. There are cases, however, when a less elaborate scheme is required and the 723 is capable of delivering an over and undervoltage monitoring circuit. Figure 5-20 shows a 723 used as an undervoltage level monitor. The V_{REF} of the 723 has its voltage divided across R4 to yield 2.0 V. The voltage to be monitored is divided down by R_M and R1 and applied to the non-inverting (+) input of the error amplifier. The voltages across R1 and R4 are compared with one another. When the input voltage (V_M) is in its nominal state, the output of the error amplifier is high or 3.3 V by clamping or limiting the COMP terminal to the junction of R3 and R2 on the V_{REF} voltage divider. If the monitored input voltage (V_M) should fall below its nominal value, the amplifier changes states and goes low. Resistor R6 can be used in this instance to drive one TTL load (1.6 mA at 0.4 V_{MAX}). This can be used to set off an alarm, indicating that the voltage has fallen below an acceptable level. The monitoring of overvoltages can be accomplished in precisely the same manner; the only modification needed is that the two inputs to the error amplifier be reversed as shown by the dotted lines in the illustration.

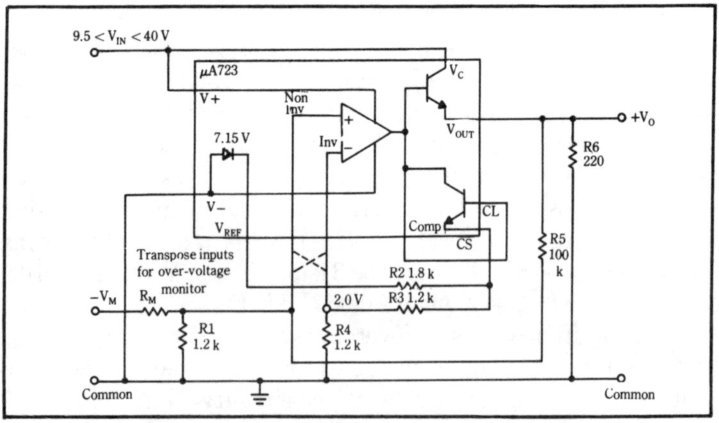

Fig. 5-20. A 723 undervoltage level monitor. (From *National Semiconductor Voltage Regulator Handbook*, coypright 1982 by National Semiconductor; reprinted with permission.)

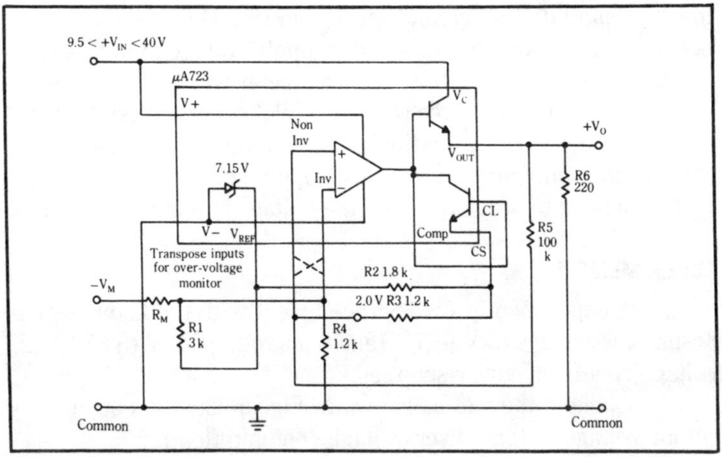

Fig. 5-21. A 723 negative under-voltage level monitor. (From *National Semiconductor Voltage Regulator Handbook,* copyright 1982 by National Semiconductor; reprinted with permission.)

Figure 5-21 shows monitoring voltages from a negative supply. The monitored voltages are referenced to V_{REF} to provide level shifting from any negative input to the +2.0 V threshold voltage. Response times for both this circuit and the one in Fig. 5-20 are typically one microsecond or less.

EXPERIMENT 5-1

This experiment examines the three internal voltage references possible with the 723 voltage regulator. The first voltage reference is a temperature-compensated zener. Another reference is a zener derived by back-biasing the emitter-to-base junction of the current-limiting transistor. The last voltage reference is a zener from the output voltage pin (10), cathode, to the V_Z pin (9), anode.

Note: A DIP package 723 must be used for this experiment because the 10-pin TO-100 "tin can" does not have enough pins to bring out all of the functions described above.

1. Connect the 723 as shown in Fig. 5-23 initially with only the power and ground connections made.

2. Measure the output voltage on the reference zener diode (pin 6). This should be approximately 6 to 7 volts.

3. Vary the voltage until the zener diode voltage just begins to drop. Note this voltage and determine the difference in the supply voltage and the zener voltage.

4. Apply a 15 V source with a 4.7 kilohm resistor in series to

the V_{OUT} pin (10) and ground the V_Z pin (9). Measure the voltage between these two pins. Lower the supply voltage until this zener voltage begins to drop. Note the voltage at which this occurs.

5. Apply a 15 V source with a 4.7 kilohm series resistor to the emitter of the current-limiting transistor (pin 3) and ground the current-limit pin (pin 2). Lower the supply voltage until this zener voltage begins to drop out. Note the voltage at which this occurs.

EXPERIMENT 5-2

This experiment provides the student with the knowledge to design low-voltage (2 V to 7 V) and high-voltage (7 V to 35 V) 723 voltage regulator power supplies.

1. Connect the 723 as shown in Fig. 5-22. The equation for output voltage in this lower voltage configuration is:

$$V_0 = \left(\frac{R2}{R1 + R2}\right)(V_{REF}) \qquad \textbf{Equation 5-7}$$

where V_{REF} = reference voltage

You have already determined the exact value for V_{REF} in Experiment 5-1. The sum of R1 + R2 must be greater than 1.5 kΩ.

2. Measure the values of R1 and R2 with a DMM (Digital

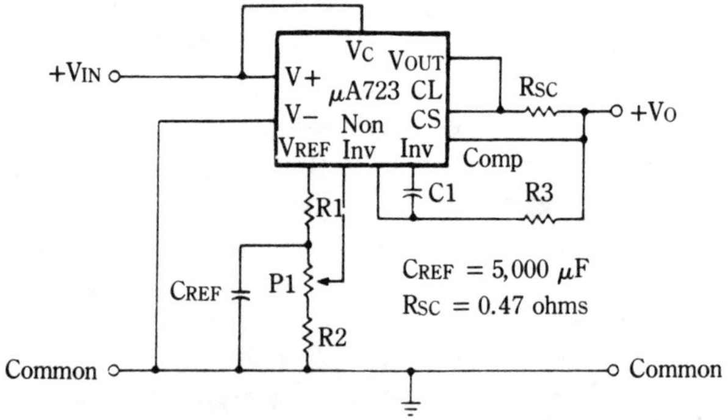

Typical performance	
Regulated output voltage	5 V
Line regulation (ΔV_{IN} = 3 V)	0.5 mV
Load regulation (ΔI_L = 50 mA)	1.5 mV

Note: $R3 = \frac{R1 \cdot R2}{R1 + R2}$ for minimum temperature drift

Fig. 5-22. 723 setup for experiment 5-2 step 1.

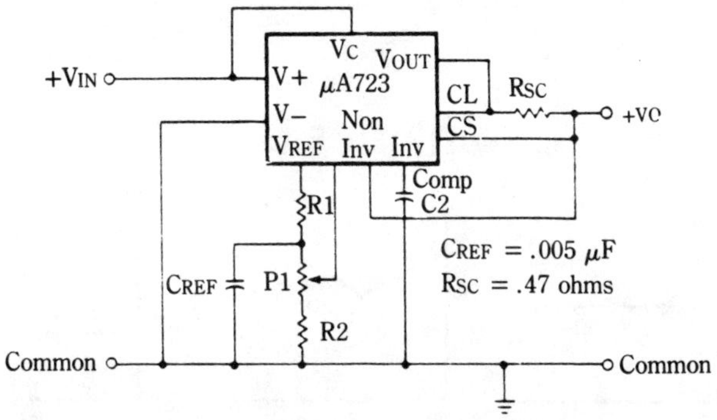

Typical performance

Regulated output voltage	5 V
Line regulation (ΔV_{IN} = 3 V)	0.5 mV
Load regulation (ΔI_L = 50 mA)	1.5 mV

Fig. 5-23. 723 setup for experiment 5-2 step 3.

Multimeter) and see how closely the output voltage adheres to the equation for output voltage. Change values of R1 and R2 and see how closely the equation describes the output voltage.

3. Connect the 723 as shown in Fig. 5-23. The output voltage in this higher voltage application is:

$$V_0 = \left(\frac{R1 + R2}{R2}\right)(V_{REF}) \qquad \textbf{Equation 5-8}$$

where V_{REF} = reference voltage

4. Substitute the various values of resistors shown for R1 and R2 in Table 5-1 and repeat step 2.

Table 5-1. Resistor Values for Experiment 5-2.

Output Voltage	R1	R2
3.0 V	10 kΩ	9.375 kΩ
4.0 V	10 kΩ	18.18 kΩ
5.0 V	10 kΩ	41.67 kΩ
6.0 V	667 Ω	20.0 kΩ
8.0 V	10 kΩ	34.44 kΩ
10.0 V	10 kΩ	16.32 kΩ
12.0 V	10 kΩ	10.69 kΩ
15.0 V	10 kΩ	7.05 kΩ
18.0 V	10 kΩ	5.25 kΩ
24.0 V	10 kΩ	3.48 kΩ

This table assumes V_{REF} = 6.2 V

Chapter 6

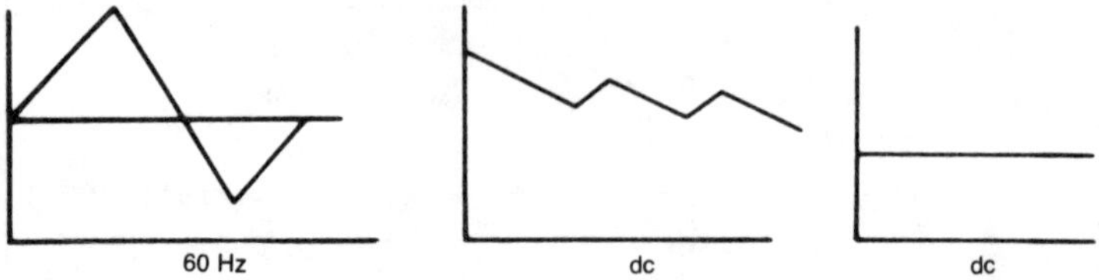

Monitoring the Power Supply's Health

This chapter examines in detail techniques and circuits to monitor critical power supply operating parameters such as over/undervoltage, overcurrent, controlling power-up rise time, thermal shutdown, etc. These generalized methods are first explained through simple single discrete op-amp designs. This is followed by more elaborate and accurate visual continuously indicating circuits. Finally, a whole power supply monitoring subsystem on a single integrated circuit, the AM6300, is examined along with the MC3424 under/overvoltage protection IC. Design examples and applications of each of these more complex ICs are included. This chapter then progresses from the simple to the more complex circuits.

DETECTING VOLTAGE LEVELS

The simplest way to detect a voltage level is with a voltage comparator (op amp using no feedback). Refer to Fig. 6-1 and note that the noninverting input (+ input) has a voltage on it, called the reference voltage. If the inverting input (– input) has a lower voltage on it, the output will go low or approach the potential of the negative supply. If the voltage on the noninverting input (– input) has a voltage greater than this reference voltage, the output will go high or approach the potential of the positive supply. Typically, in a 741 bipolar op amp, these levels come to within 1.0 to 1.5 volts of the two supply voltages.

There are times when you will want to monitor a narrow range or "window" of voltages with upper and lower limits. The circuit in Fig. 6-2 accomplishes this. This circuit has its lower and upper

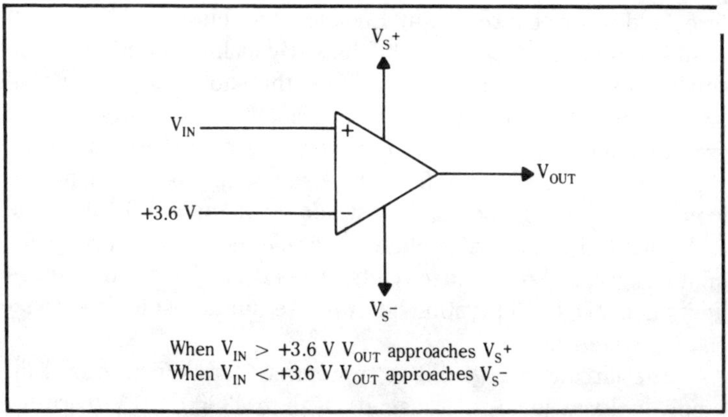

Fig. 6-1. A basic voltage comparator.

thresholds determined by a single zener diode and the width of the window determined by the ratio of two resistors, R1 and R2. The lower triggering point occurs when V_1 is greater than V_2 and Equation 1 within the illustration is met. The comparator turns off (the upper triggering point) when Equation 2 is met. The window width is expressed by Equation 3 in Fig. 6-2. If V_{IN} becomes larger than $1.2\ V + V_Z$, current is limited through diode D2 so that the lower

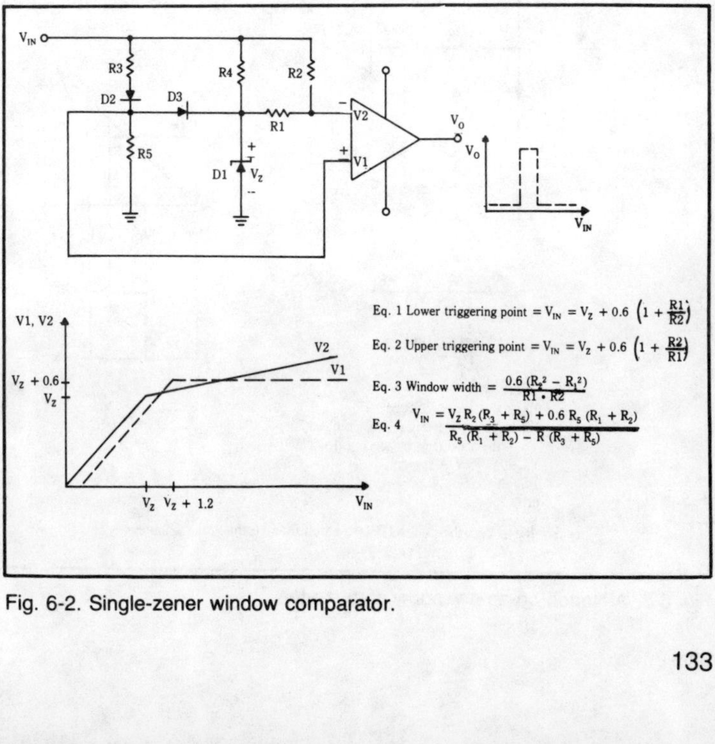

Fig. 6-2. Single-zener window comparator.

threshold is now expressed by Equation 4 within Fig. 6-2. Typical values of R4 and R5 can be 10 kilohms. By making the R3/R5 ratio smaller, the two equations for the lower threshold trip point (Equations 1 and 4) become closer to one another, even when V_{IN} is greater than V_Z plus 1.2 V. However, this circuit would not be appropriate to monitor a +5.0 Vdc power supply powering a TTL logic printed circuit board. You would want the +5.0 Vdc to be within a ±0.5 V range for commercial grade and a ±0.25 V range for military grade integrated circuits. Operating TTL above these limits is damaging; operation below these limits results in erroneous logic states.

The circuit in Fig. 6-3 is appropriate for detection of TTL supply voltage levels because it is an LM3900 CDA (current-differencing amplifier) type quad op amp. The LM3900 and MC3401 are two of the most popular CDA op amps. Their main advantages are that they operate from a single-ended power supply and go down to below +5 V with little change in open-loop characteristics. Many

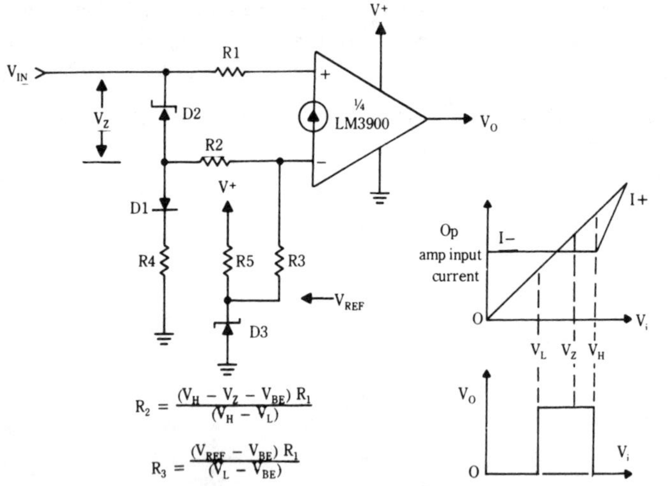

$$R_2 = \frac{(V_H - V_Z - V_{BE})\,R_1}{(V_H - V_L)}$$

$$R_3 = \frac{(V_{REF} - V_{BE})\,R_1}{(V_L - V_{BE})}$$

where
V_Z = the zener voltage
V_H = the upper triggering point
V_L = the lowering triggering point
$V_{BE} \simeq$ 0.65 V A typical silicon diode voltage drop.

Set R5 to bias the zener D3 into a point above the zener's knee.

Fig. 6-3. A Norton op-amp window comparator.

bipolar op amps do not work at all below +5 V, making them unsuitable for this type of application.

CDA op amps, as their name suggests work on the principle that the input amplifier to the op amp experiences a difference in input currents, $\left(I_{IN}\ (+)\ \text{minus}\ I_{IN}\ (-)\right)$ but with negative feedback (when the loop is closed) this difference in current is theoretically made to equal zero. Also, when the loop is closed, the current into the (−) input is made to equal the current into the (+) input. You may recall that this is precisely how the voltages of a regular op amp perform. This is the main difference between these two types of op amps. Note the circle with the arrow inside it placed between the two inputs to the op amp. This immediately identifies a CDA, or Norton op amp as they are sometimes called. These operational amplifiers are covered in Chapter 9 of *The Design of Operational Amplifier Circuits with Experiments* by Howard M. Berlin, E & L Instruments, copyright 1977.

The circuit in Fig. 6-3 will give an indication in either the usual positive direction, as shown in the illustration, or in the negative direction if the inputs are reversed. The input voltage monitored can also be well above the supply voltage to the Norton op amp. The value of the zener voltage of diode D2 (V_Z) must be low enough to allow the current through R2 to be 3 mA or less. Refer to the two equations within Fig. 6-3, and start by selecting R1 to give a current between 10 microamps up to the previously mentioned 3 mA limit. The selection of R2 and R3 can then follow. The second zener, D3, along with R5 establish the reference voltage, V_{REF}, in the second equation. If a "sharp-kneed" zener is used, such as an LM103 reference diode, then D1 and R4 may be eliminated. The LM103 comes in 13 different values from 1.8 to 5.6 volts and can be used here to replace the 4.7 volt zener, D2. The knee is truly sharp with just a 15 mV typical change over a 100 μA to 1 mA biasing current range.

In Chapter 8, there is a section on fans and blowers used to vent hot air away from heat-producing elements within an equipment cabinet. To conserve power, it would be advisable to have a temperature monitor as shown in Fig. 6-4, so that fans do not constantly run. Note from the illustration (Fig. 6-5) that three temperatures, T1, T2, and T3, are distinct switching points. T1 (35° C) causes the fans to come on, T2 (40° C) turns on an overtemperature alarm and T3 (45° C) turns off the equipment. Three of the four op amps are used as comparators with 5° C of hysteresis each. The remaining op amp has an LM335 thermal-sensor IC producing a calibrated output

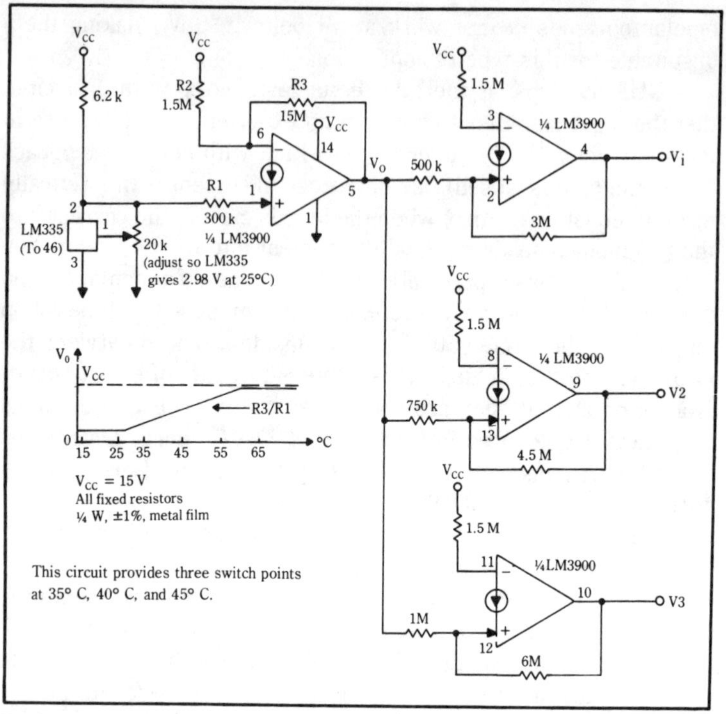

Fig. 6-4. A temperature sensor window comparator.

voltage in proportion to temperature. The design equations for the LM335 are shown in Fig. 6-5. This is an application of the bandgap principle in which the extremely linear relationship of V_{BE} vs. temperature is used. The LM335 bandgap thermal sensor has been calibrated here for a 2.98 volt output at 25° C with this voltage fed to the other three op amps.

CONTINUOUSLY MONITORING VISUAL INDICATORS

There are applications in which it is mandatory to continuously monitor one or more voltage levels and to have a visual indication of the status of these monitored voltages. The following section shows seven such circuits, all monitoring voltages with visual representation. One monitors the presence as well as the level of TTL logic signals.

The first circuit, Fig. 6-6, shows how an LM339 quad op-amp comparator arrangement monitors +5 V, +12 V, and −12 V power supply outputs. Each of these comparators compares the outputs of these three power supplies with a reference voltage from a resistor network from a +20 V source (V2). The +5 V monitor is set at 6.1 V and when this is reached, it energizes an LED and causes the beeper to sound. Simultaneously, pin 5 drops to about +5.2 V. This creates a hysteresis band between +5.2 V to +6.1 V so that possible oscillations during transition are minimized. The second com-

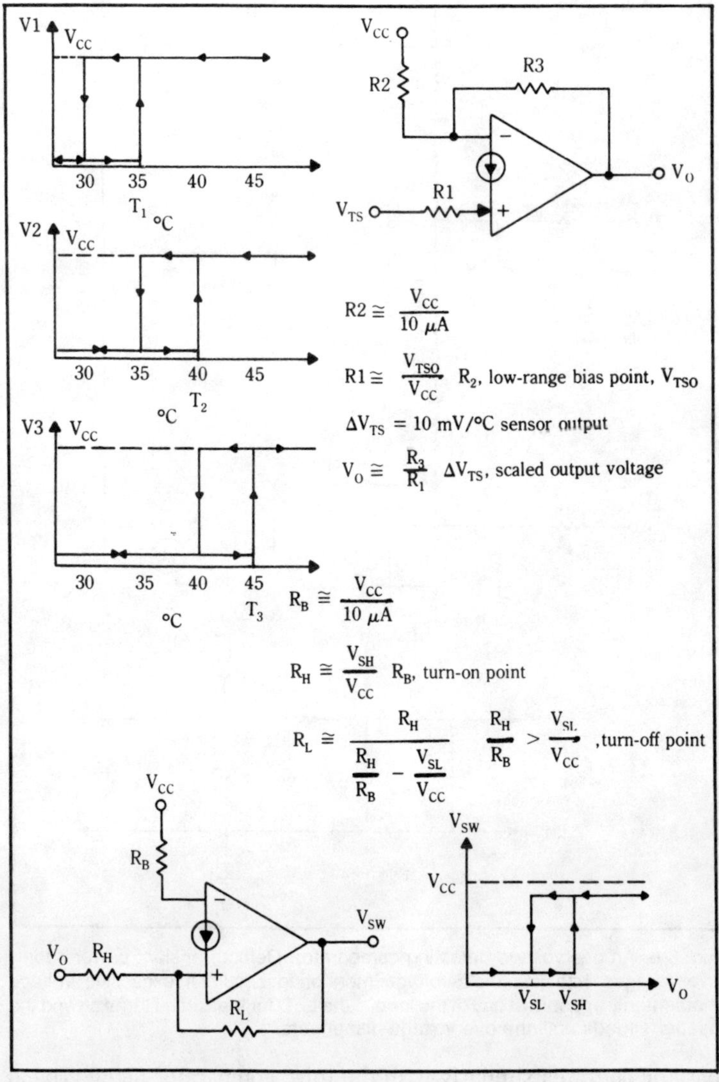

Fig. 6-5. Approximate design equations for the range-scaling amplifier (top) and comparator (bottom) portions of Fig. 6-5 are used to select the values of all the required metal-film resistors.

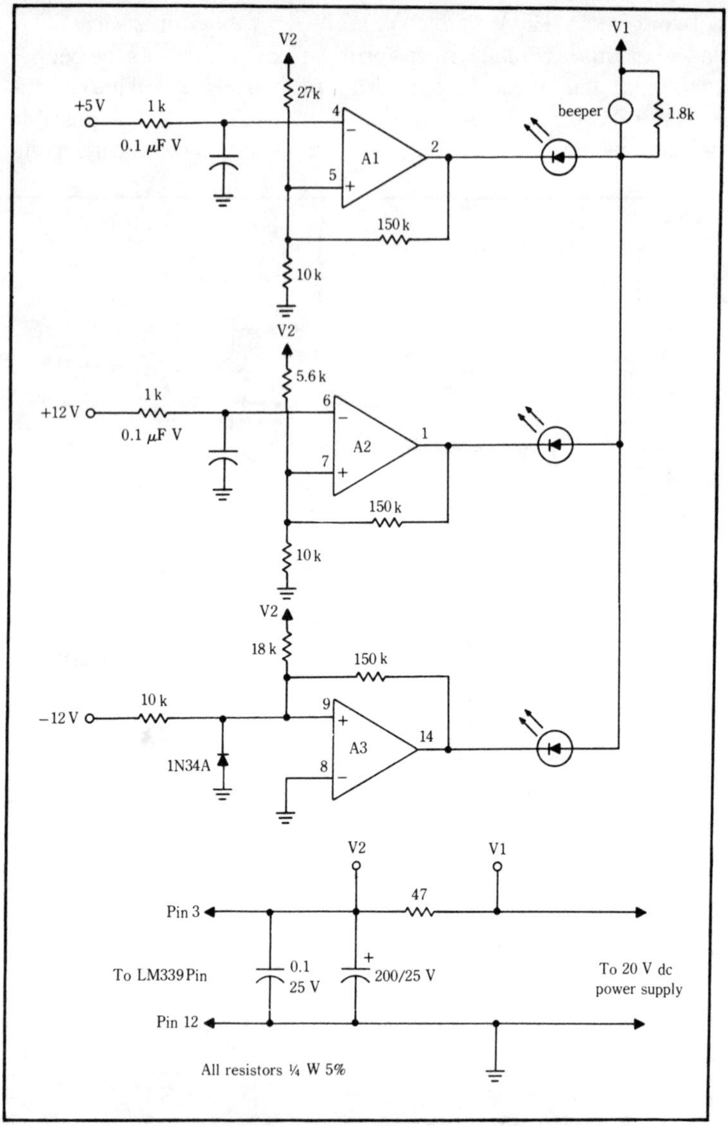

Fig. 6-6. An overvoltage detecting comparator. Detect transient power-supply overvoltages with this triple-voltage monitoring circuit. If excessive voltage momentarily appears at any of the inputs, the LED for that circuit turns on and the beeper sounds until the overvoltage disappears.

parator operates similarly with the third comparator using ground as its reference level. The monitored −12 V is balanced against the +20 V comparator supply and when the comparator senses a voltage

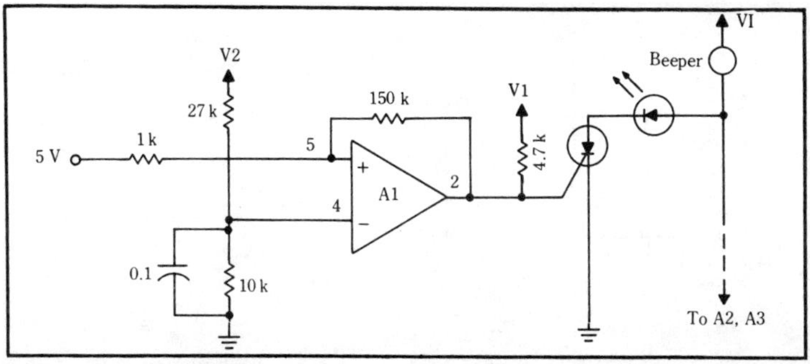

Fig. 6-7. A latching overvoltage detecting comparator. Maintain an alarm condition when the excessive voltage disappears by tying each comparator output to an SCR gate. Note that the comparator inputs are reversed when using an SCR.

above −12 V, the comparator's output switches and sounds the alarm. If you want to leave this monitor unattended and wish to know if an overvoltage occurred, a latching arrangement, such as the SCR shown in Fig. 6-7, may be obtained with the fourth unused op amp within the pack.

Dual LEDs

Light-emitting diodes are available in single, two, and three lead packages with two single LEDs contained therein. Each can also be a different color (Fig. 6-8). These colors are usually red and green. The colors visible to the human eye have a spectrum of red, orange, yellow, green, etc, ; therefore, with a red and green LED, you have the ability to produce the colors of orange and yellow as well which are contained between red and green in the visible light spectrum. This "mixing" of composite colors proves to be very useful in voltage sensing and indicating circuits. More specifically, if lead 2 is more positive than lead 1, (refer to Fig. 6-8), the green LED will light. Conversely, if lead 1 is more positive than lead 2, the red LED will light. If the two LEDs are driven by a square wave (refer to Fig. 6-9) that alternately makes one LED more than the other, the color of this "favored" LED will naturally predominate. Therefore, a composite of both colors, as well as the distinct colors of red and green themselves, and varying shades in between are all

Fig. 6-8. Dual LEDs.

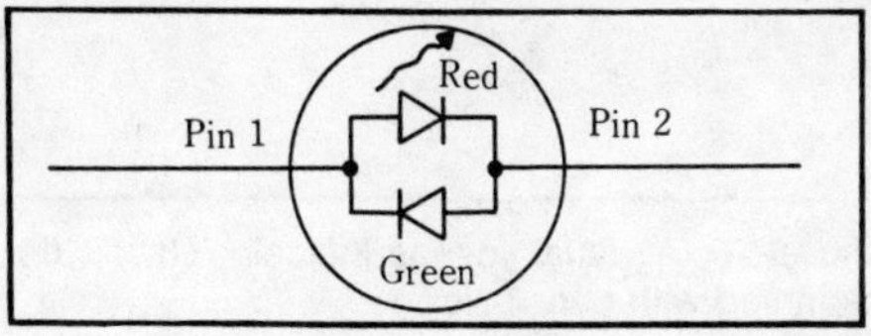

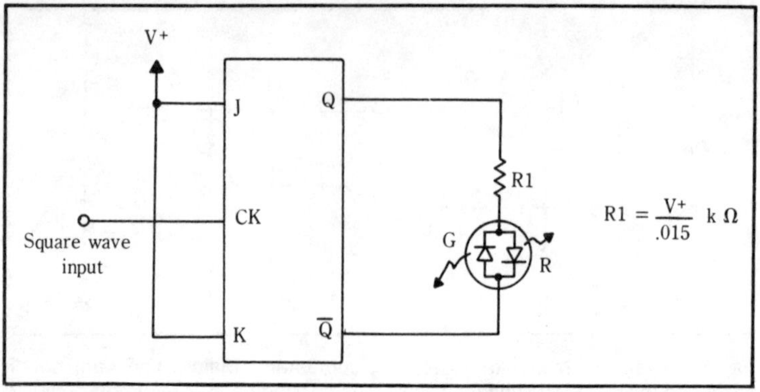

Fig. 6-9. A dual-LED driver circuit.

possible with a dual LED and a circuit capable of producing square waves of variable duty cycles.

Voltage Indicating Circuits Using Dual LEDs

The circuit in Fig. 6-10 uses the principles set forth in the preceding paragraph. The circuit uses a quad LM324 op amp with amplifier A1 (see Fig. 6-11) the input which can interface with, for example, a voltage follower. Its output, V_i, is applied to A2 which is a voltage-controlled pulse-width modulation amplifier; Fig. 6-12 shows its input voltage vs. pulse-width ratio. It varies as the

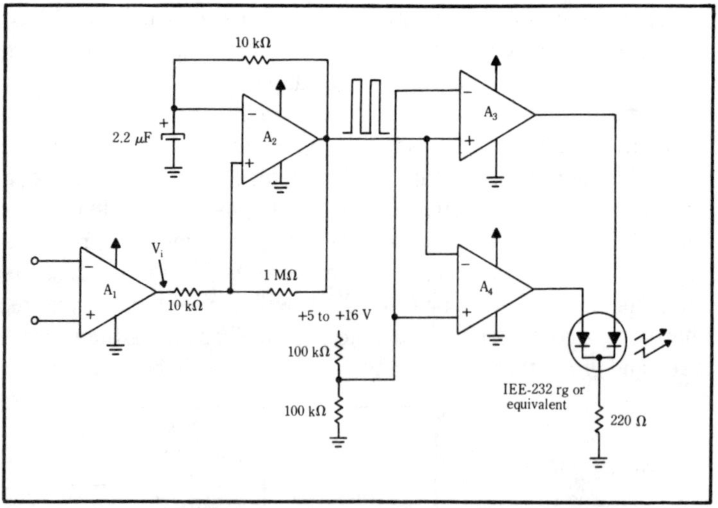

Fig. 6-10. A visual voltage indicator circuit. (From *Electronics*, April 7, 1981, reprinted with permission.)

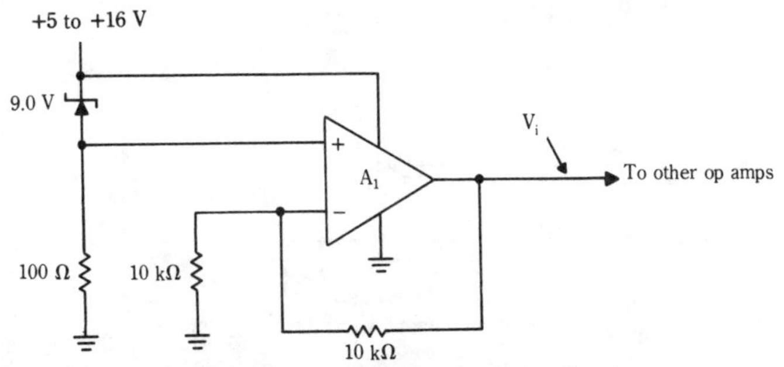

B^+	V_i	V_i/B	PW_H/PW_L	Color
13.0	8.0	0.62	2.00	Green
12.5	7.0	0.56	1.25	Green/yellow
12.0	6.0	0.50	1.00	Yellow
11.5	5.0	0.43	0.60	Orange
11.0	4.0	0.36	0.40	Red

Fig. 6-11. "Watcher" circuit.

exponent of V_i over two decades. The output of A2 is applied to A3 and A4, noninverting and inverting amplifiers, respectively.

The amplifiers power the dual LED so that when the red LED is on, the green LED is off and vice-versa. The table in Fig. 6-11 indicates color vs. pulse-width ratios and duty cycles. This table

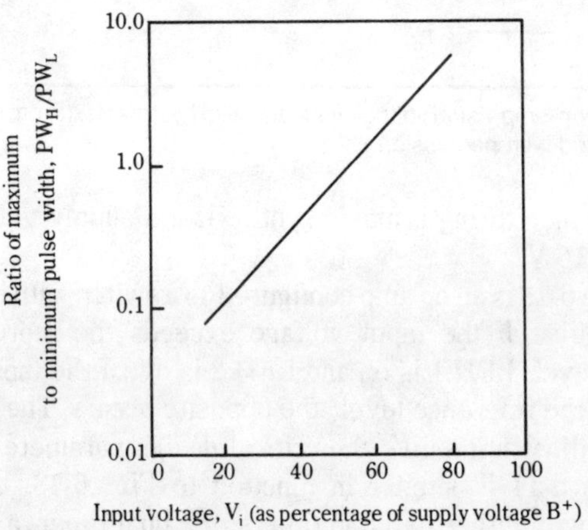

Fig. 6-12. Input voltage vs pulse-width ratio. (From *Electronics*, April 7, 1981; reprinted with permission.)

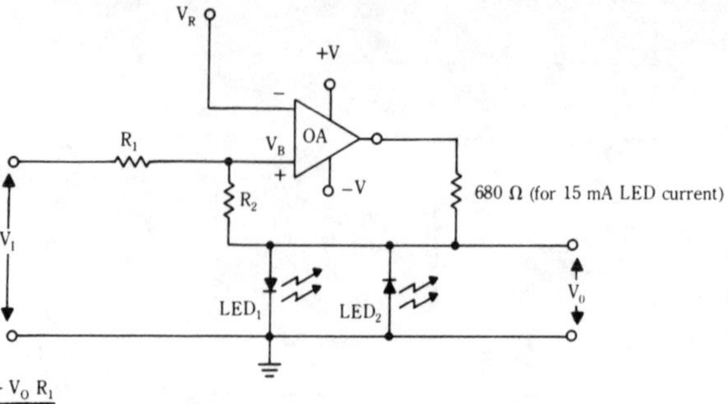

$$V_B = \frac{V_I R_2 + V_O R_1}{R_1 + R_2}$$

At the moment V_O changes from $+V_{D1}$ to $-V_{D2}$, the following situation exists:

$$V_R = V_B = \frac{(V_{I1} R_2) + (V_{D1} R_1)}{R_1 + R_2}$$

When V_O changes from $-V_{D2}$ to $+V_{D1}$ the situation is:

$$V_R = V_B = \frac{(V_{I2} R_2) - (V_{D2} R_1)}{R_1 + R_7}$$

In both cases V_R is unchanged, hence:

$$\frac{R_1}{R_2} = \frac{V_{I2} - V_{I1}}{V_{D1} + V_{D2}}$$

where V_{D1} and V_{D2} are the LED voltages. $V_{I2} - V_{I1}$ is the desired range (note $V_{I1} < V_{I2}$ to maintain a positive ratio). V_R can be calculated from Eq. 1. For example, if $V_{D1} = V_{D2} = 1.6$ V (not necessarily equal if green and red LED's are used) the chosen trip levels are:

$V_{I1} = -1.2$ V for LED_2 to switch ON;
$V_{I2} = +2.0$ V for LED_1 to switch ON;
substituting in Eq. 3 yields,

$$\frac{R_1}{R_2} = 1 \text{ (choose } R_1 = 10 \text{ k, consequently } R_2 = 10 \text{ k).}$$

V_R, the required reference voltage, can be calculated from Eq. 1 or Eq. 2.

$$V_R = \frac{-1.2 \times 10^6 + 1.6 \times 10^4}{10^4 + 10^4} = +0.2 \text{ V}$$

Fig. 6-13. An op-amp visual voltage indicator with hysteresis. (From *EDN*, May 5, 1974; reprinted with permission.)

represents monitoring a music synthesizer's supply voltage from +5 V to +16 V.

Figure 6-13 is an op amp configured as a switch with hysteresis characteristics. If the input voltage exceeds the more positive reference level, LED 1 is on and LED 2 is off. If the input voltage falls below the reference level, the opposite occurs. The equations within the illustration give the critical design parameters.

Figure 6-14 is similar in function to Fig. 6-13, but is implemented with either two 555 timers or a dual timer, a 556. This circuit, however, indicates the four possible combinations available on two lines, inputs X and Y. The presence of two logic ones causes both timers to be on alternately and the 50 percent duty cycle switching of alternate conducting paths through the dual LED causes a yellow to result.

Figure 6-15 shows how arrayed LEDs can track an incoming varying signal with lower and upper setpoint potentials. These upper and lower potentials are tapped off a 0 V to 5 V potential on two pots and applied to an eight-channel analog multiplexer, the CD4051BE. This IC is clocked by the CD4024BE counter so that all eight channels are scanned in sequence. These are converted into control signals lighting the appropriate LEDs in the matrix. The LED matrix is only a matrix by its electrical hookup. It may be physically configured as a row of 64 LEDs, as shown in the upper portion of the illustration. Also, a dual comparator may be added to sound an alarm if an incoming signal goes beyond the preset range.

The next circuit uses an LM3914 bar-graph driver IC in the expanded scale of operation to continuously monitor a very narrow window of voltages and drives a 10-LED array to monitor this window. Such a circuit is handy in grading voltage regulators for a near uniform output voltage. Before the circuit is described, the LM3914 needs to be discussed.

LM3914

The LM3914 bar-graph driver IC (refer to Fig. 6-16), drives either 10 LEDs or a 10-LED bar-graph DIP (dual inline package)

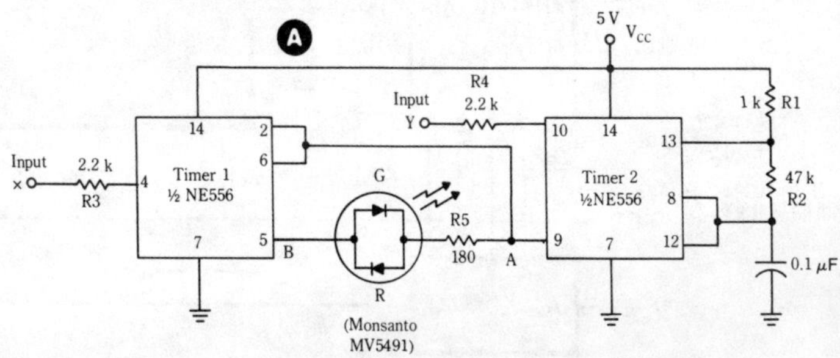

X	Y	LED color
0	0	Off
0	1	Red
1	0	Green
1	1	Yellow

B

Two LEDs in a single package, driven by an NE556 dual-timer (A) indicate four combinations of binary inputs X and Y (B). The LED emits either red, green or yellow light, or stays off.

Fig. 6-14. A dual-LED/timer logic state indicator.

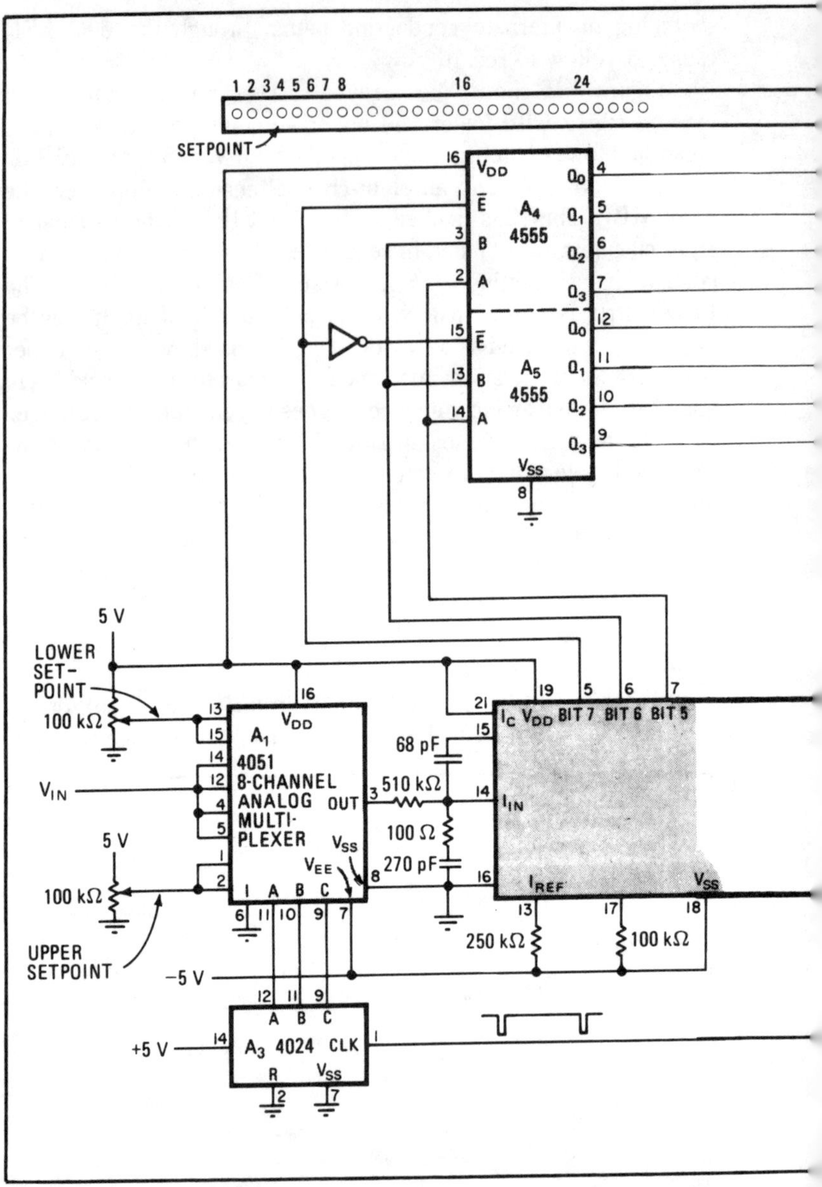

Fig. 6-15. An arrayed LED signal tracking circuit. Eight-by-eight-LED matrix and C-MOS logic form analog dual-setpoint meter. Setpoint potentials and data voltage V_{in} are each introduced into multiplexer and scanned in sequence, then converted into control signals that light the appropriate LEDs in the matrix. Setpoint LEDs are half the brightness of the LED representing V_{in} because they are sampled at half the rate. (From *Electronics*, April 26, 1979; reprinted with permission.)

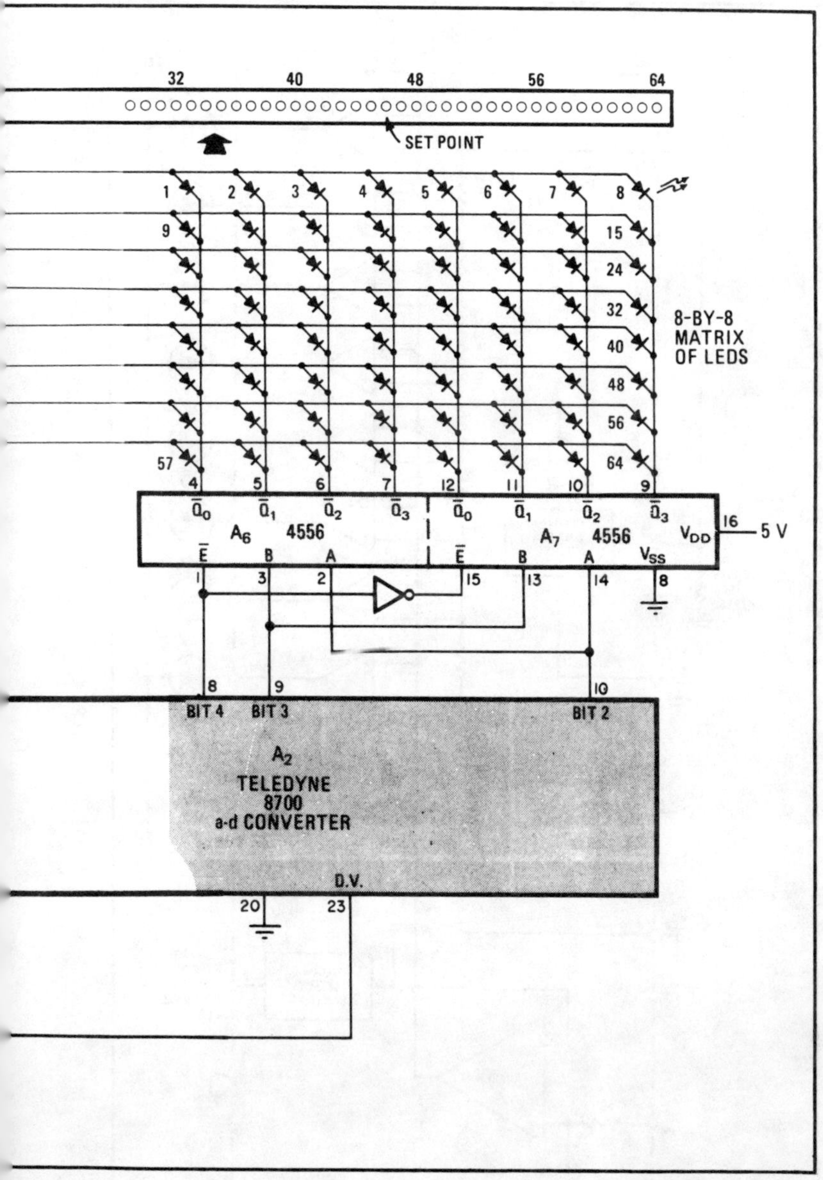
32
40
48
56
64
SET POINT
8-BY-8
MATRIX
OF LEDS
A6 4556
A7 4556
VDD
5 V
VSS
BIT 4
BIT 3
BIT 2
A2
TELEDYNE
8700
a-d CONVERTER
D.V.

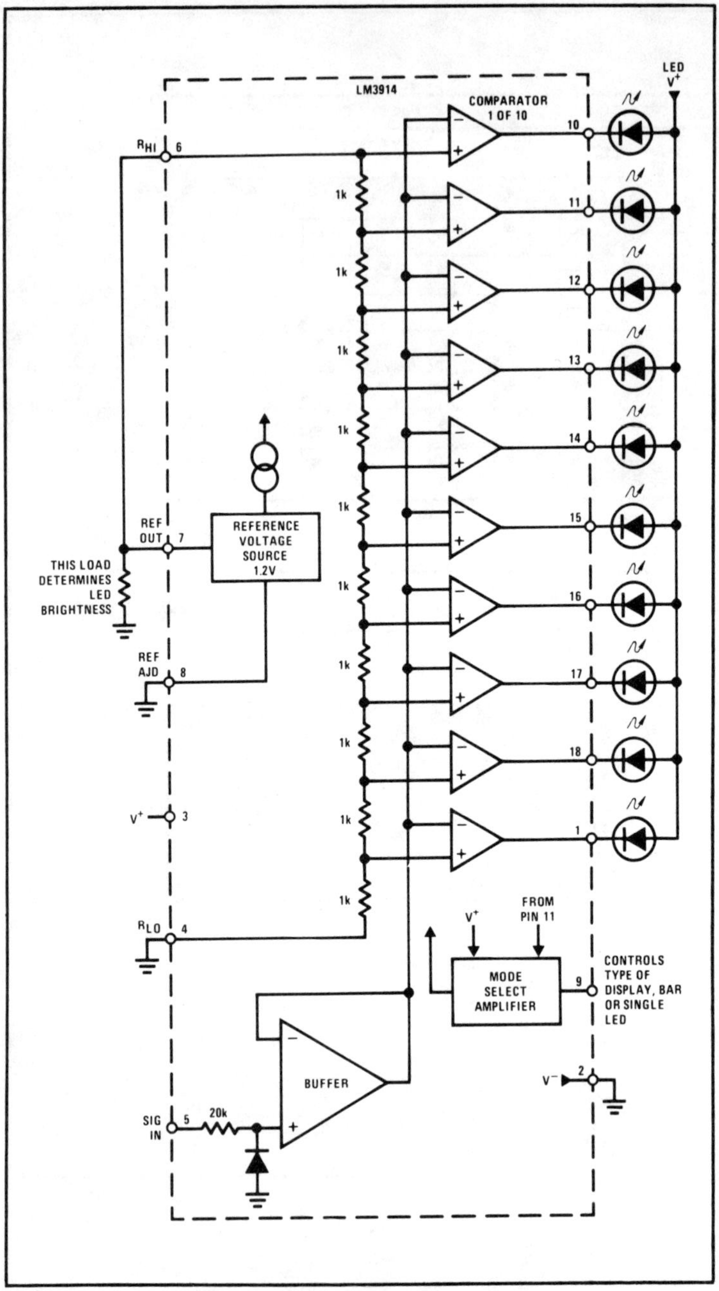

Fig. 6-16. The LM3914 schematic block diagram. (Copyright 1978 by National Semiconductor; reprinted with permission.)

display (refer to Fig. 6-17). The display can be driven in either the bar-graph or dot modes. The bar-graph mode is activated if pin 9 (mode select) is tied to a potential of within 200 mV of V^+, the supply voltage of the IC. Anything less activates the dot mode. Therefore, tying pin 9 to pin 3 ensures bar-graph operation. Tying pin 9 to pin 11 (the next to top LED driver output) ensures dot-mode selection.

The dot-mode of operation entails moving a dot (single bar-graph segment) up the display. The bar-graph mode, as its name implies, acts like a bar graph with all the segments lighting—right up to the level of the last driven segment. As an example, in the bar-graph mode, if the LM3914 is configured to detect a 5 V span, each of the ten bar segments represents 0.5 V, and if 3.5 V is present on the input, seven segments will light.

The driver outputs of the LM3914 are current-regulated open-collector transistors with programmable currents from 2 mA to 30 mA and with interface capability to both TTL and CMOS logic. The brightness of the LEDs is determined by "programming" the output current through a resistor on pin 7. The current flowing through the resistor on pin 7 has a 1.2 V reference supplying this current. The current from pin 7 will be approximately one-tenth that of the output. Therefore, if a 1.2 kilohm resistor is placed from 7 to ground, a 1 mA current flows through it, but a 10 mA current flows through the LEDs on the output. This "programming" current capability allows the elimination of the traditional series-limiting LED resistors. A typical red LED can safely withstand 16 mA to 18 mA.

Refer to the LM3914 internal block diagram, Fig. 6-16, and note the presence of ten comparators with their noninverting (+) inputs all tied to a voltage divider chain. The ten resistors are identical so equal voltage division occurs on each comparator. Specifically, if the reference high (R_{HI}—pin 6) is held at 7.5 V and

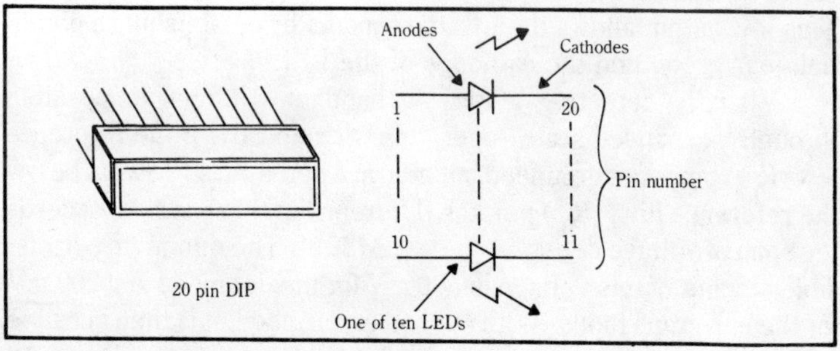

Fig. 6-17. A ten-segment bar-graph display.

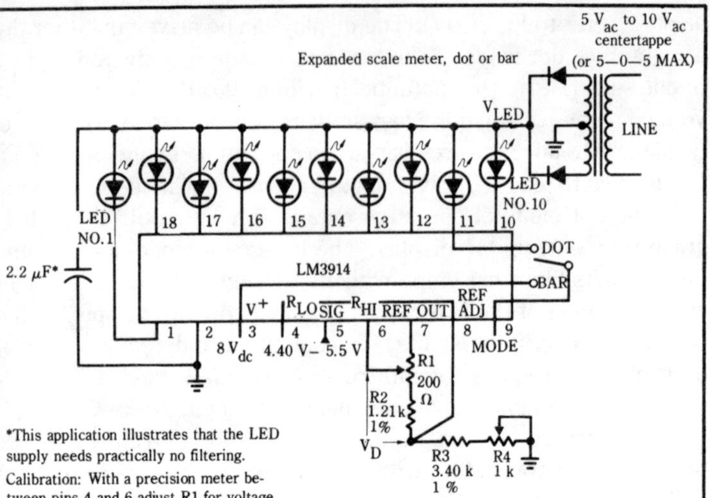

*This application illustrates that the LED supply needs practically no filtering.

Calibration: With a precision meter between pins 4 and 6 adjust R1 for voltage V_D of 1,20V. Apply 4.94 V to pin 5, and adjust R4 until LED No. 5 just lights. The adjustments are non-interacting.

Application Example: Grading 5 V Regulation

HIGHEST NO. LED ON	COLOR	$V_{OUT(MIN)}$
10	Red	5.54
9	Red	5.47
8	Yellow	5.30
7	Green	5.18
6	Green	5.06
	5V	
5	Green	4.94
4	Green	4.82
3	Yellow	4.7
2	Red	4.58
1	Red	4.46

Fig. 6-18. A 5 V voltage regulator grading circuit. (Courtesy National Semiconductor.)

the reference low (R_{LO}—pin 4) is grounded, a voltage of 0.75 V appears at the input of the first or "bottom" comparator, 1.5 V appears on the second comparator, 2.25 V appears on the third comparator, etc. Each comparator with a voltage on it becomes active by having the output of that particular comparator go low. This low output allows the LED to conduct by establishing a potential near ground on the cathodes of the LEDs.

Visually detecting a narrow band of voltages is possible through "expanded scale" operation. Specifically, if the reference low (R_{LO}) pin is not grounded but held at a potential somewhat below the reference high (R_{HI}) pin, the difference in these two voltages is the span of voltage detected by the LM3914. The minimum permissible amount of this voltage is 500 mV for the dot mode and 200 mV for the bar-graph mode. With the bar-graph mode, it is then possible to detect increments as small as 20 mV. The circuit in Fig. 6-18 detects 120 mV increments because V_D is ten times that amount, or 1.2 volts. A "finer" grading of regulators could be accomplished by lowering the value of voltage across R_{HI} to R_{LO} to as low as 200 mV in the bar-graph mode. Note from the table within the illustration that the detected increments are in steps of 120 mV. The circuit in Fig. 6-18 uses green LEDs to indicate an acceptable range, yellow LEDs to indicate a marginal range, and red LEDs to indicate an unacceptable range. Discrete LEDs may be used in lieu of bar-graph DIP displays, as is done here, with an information-conveying visual effect. However, remember that the single resistor from pin 7 on the LM3914 is the sole determining factor in the magnitude of output currents for all of the current drivers, so all LEDs have the same current. However, different colored LEDs have varying brightnesses per given current. Green is the least bright, while red is the brightest. Specifically, a red MV5774, LED at 20 mA has 5 mcd (microcandela) of brightness, an identical yellow, MV5374 LED has 4.0 mcd, and a green, MV5174 LED has only 1.8 mcd.

The circuit in Fig. 6-19 detects and displays dc voltages with high resolution down to 20 mV increments. The circuit has programmable upper and lower limits as a result of using the expanded scale of operation of the LM3914. If the input voltage is overrange, the dual LED will be red, underscale indicates green, and within range is an orange. Also, when within range, the bar-graph LED flashes in an "exclamation point" manner. Specifically, the lowest bar graph segment is "hardwired" on and serves as a power-on indicator. The remaining bar-graph segments above this flash and indicate magnitude.

Detection of upper and lower limits is implemented with a CD 4048BE programmable gate IC. This unique IC has eight inputs, one output, and three control lines. These three control lines select eight possible logic configurations (refer to Table 6-1). The eight inputs to this gate come from the first through the eighth outputs of the LM3914. When a value below the expanded-scale detecting window is applied to the input, the output is all logic zeros. Conversely, when a voltage above the window is present the output is all logic highs. But when all highs or lows are detected, we desire a logic low output. This is an exclusive-NOR function. Referring to Table 6-1 shows no such function exists. We can implement this logic function by making the CD4048BE programmable gate into an OR to detect all lows and then into a NAND to detect all highs with a resulting low output.

Refer to Table 6-1 and note how this can be done by manip-

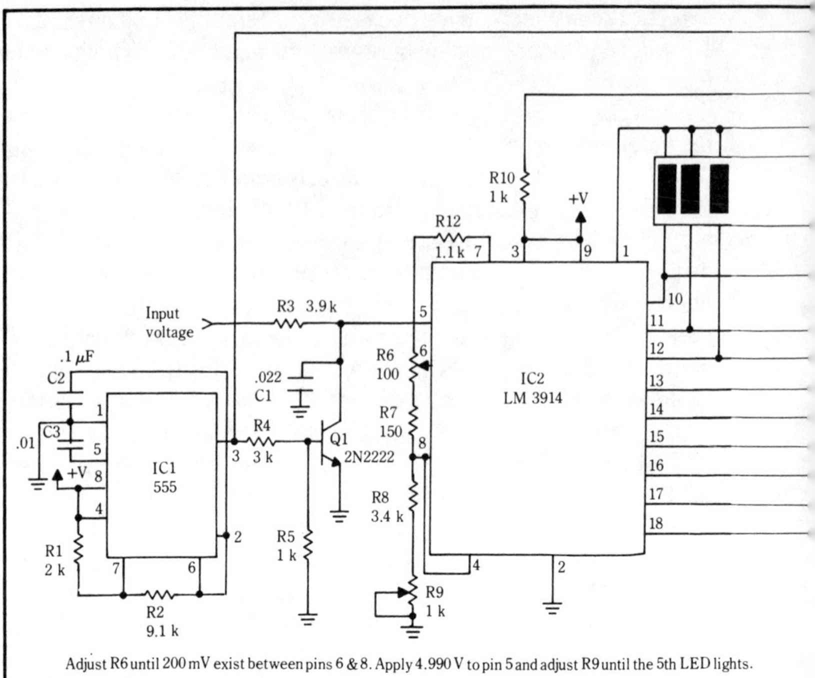

Fig. 6-19. A high resolution voltage detection circuit.

ulating the K_a line (pin 10). An OR is a 001 and a NAND is a 101; therefore, manipulation of this first bit is all that is needed. This is done by attaching pin 10 to any of the LM3914 output pins. When all logic highs occur, a NAND will be in effect and a low output results. When all logic lows are on the output of the LM3914 (underrange), then an OR logic function is in effect and again a low results from the CD4048BE's output. Using the above method, we obtain the desired exclusive-NOR function.

The visual monitoring LEDs will thus work as follows. A low causes the JK flip-flop to operate asynchronously with a high or overvoltage turning on the RED LED. An undervoltage turns on the GREEN LED. A voltage within range will have some combination

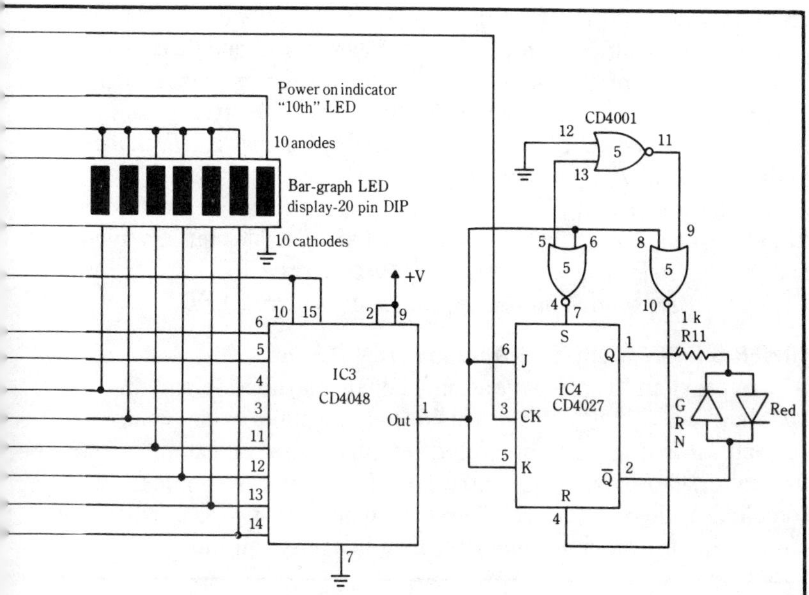

Highest No LED on	V_{OUT}
10	5.09 ← Used as on indicator
9	5.07
8	5.05
7	5.03
6	5.01
5	4.99
4	4.97
3	4.95
2	4.93
1	4.91

of logic highs and lows so no low will result from the "exclusive-NOR" programmable gate (the CD4048BE). When this condition is in effect the JK inputs are held high and the NOR gates (5A and 5B) are disabled. This allows the JK flip-flop to operate synchronously

Table 6-1. The CD4048 Truth Table.

Logic Function	K_a Pin	K_b Pin	K_c Pin	Unused Input Pin 15
NOR	0	0	0	0
OR	0	0	1	0
OR/AND	0	1	0	0
OR/NAND	0	1	1	0
AND	1	0	0	1
NAND	1	0	1	1
AND/NOR	1	1	0	1
AND/OR	1	1	1	1

and it is driven by the approximately 10 percent duty cycle astable multivibrator made from the 555 timer. Resistors R1 and R2 control the duty cycle of the output. These two resistors, along with capacitor C1, control the output frequency. A set of very handy nomographs exist on page 3-3 of the *The 555 Timer Applications Sourcebook, with Experiments*, by Howard Berlin, copyright 1976 by E&L Instruments, to enable quick and accurate astable multivibrator designs. The JK flip-flop takes this 1 kHz astable square wave and divides it in half and alternately turns on the RED and then GREEN LEDs to give an orange composite colored LED.

POWER SUPPLY MONITORING SUBSYSTEM ICS

In previous chapters we examined current-limiting and thermal shutdown circuitry that was inside the integrated circuit voltage regulator. Most modern integrated circuit voltage regulators have this internal protection circuitry built-in; however, for a truly complex and high performance piece of equipment such as a critical life-sustaining or monitoring instrument in a hospital, this internal

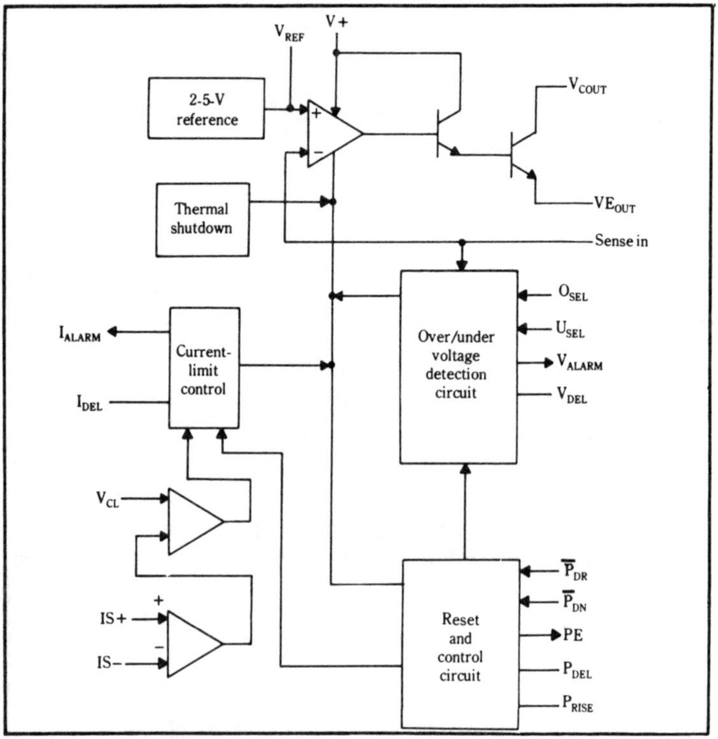

Fig. 6-20. The AM6300 power control subsystem.

protection circuitry is often not good enough. For this important job, we need a power supply system with even more sophisticated capabilities.

AM6300 Power Control Subsystem

The Advanced Micro Devices AM6300 Power Control Subsystem offers these capabilities. This 20-pin DIP IC, consisting of approximately 200 transistors, is more than a mere voltage regulator with internal voltage reference, error amplifier, series-pass power transistor, and the usual thermal and current-limiting protection circuitry. The AM6300 additionally has programmable over/undervoltage detection circuitry, a controllable variable time delay before the device shuts down in response to an extreme voltage or current condition and programmable power-down, power-up and thermal shutdown circuitry. There are also "flags" or pins on the integrated circuit which become active in one logic state and remain dormant in the opposite logic state. More specifically, if you program the overvoltage, overcurrent, and power-up conditions for specific levels and if any of these levels are exceeded, then the respective flag will go to a TTL level high. The specifics of programming levels of voltage and current, etc. will be shown shortly. But, first, let us take a detailed look at the function of each pin of this IC shown in Fig. 6-20.

V+ The supply voltage pin upon which a continuous voltage of 5 V to 40 V can be applied.

Gnd The ground pin.

V_{REF} Reference Voltage Output. The output of the internal, precision, 2.5 volt reference.

VE_{OUT} Output Transistor Emitter. The emitter of the regulator output Darlington.

V_{COUT} Output Transistor Collector. The collectors of the regulator output Darlington.

$SENSE_{IN}$ Voltage Sense Input. A divided-down portion of the output voltage is fed back into the regulator through this pin. The regulator adjusts the output so that this value equals the internal reference voltage, 2.5 volts. This voltage is also used in the over/undervoltage detection circuit.

O_{SEL} Overvoltage Limit Select. This input selects the overvoltage limit. A logic 1 selects a +5 percent limit and a logic 0 selects a +10 percent limit.

U_{SEL} Undervoltage Limit Select. This input selects

the undervoltage limit. A logic 1 selects a −5 percent limit and a logic 0 selects a −10 percent limit.

V_{CL} Current-Limit Voltage Input. This input is used to set the differential threshold of the current-limit detection circuit. The threshold will be equal to half of the voltage on this pin.

IS_+, IS_- Positive and Negative Current-Sense Inputs. The differential voltage across the current-sense resistor is applied to these inputs and compared to the current-limit threshold.

I_{DEL} Current-Limit Delay Control. The delay time during which the current limit can be exceeded without shutting down the regulator is proportional to the value of a capacitor placed between this pin and ground.

V_{DEL} Voltage-Limit Delay Control. The delay during which an over or undervoltage limit can be exceeded without shutting down the regulator is proportional to the value of a capacitor placed between this pin and ground.

V_{ALARM} Voltage Limit Alarm. This open-collector output goes low when the regulator shuts down due to an over or undervoltage limit being exceeded. The V_{ALARM} output is reset by taking the $\overline{P}_{DR}$ or $\overline{P}_{DN}$ input low or by removing power from the AM6300.

I_{ALARM} Current-Limit Alarm. This open-collector output goes low when the regulator shuts down due to the current limit being exceeded. The I_{ALARM} output is reset by taking the $\overline{P}_{DR}$ or $\overline{P}_{DN}$ input low or by removing power from the AM6300.

$\overline{P}_{DR}$ Power-Down Reset (With Delay). An active low input on this pin shuts the regulator down and resets the voltage and current alarms. When this input is taken high again the regulator begins to power-up after a user-programmed delay.

$\overline{P}_{DN}$ Power-Down Reset (Without Delay). An active low input on this pin shuts the regulator down and resets the voltage and current alarms. When this input is taken high again the regulator begins to power-up with no delay.

P_{DEL} Power-Up Delay Control. A capacitor between this pin and ground sets the delay which occurs after

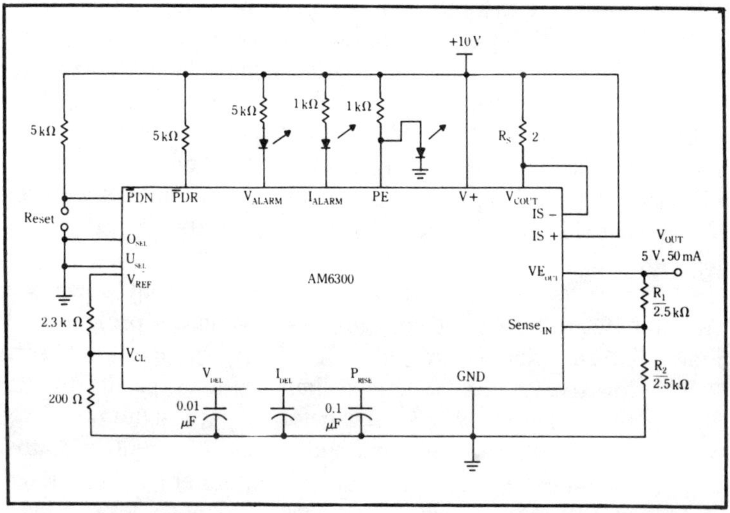

Fig. 6-21. A 5 V, 50 mA precision AM6300 based power supply.

$\overline{P}_{DR}$ is taken high and before the regulator begins to power up. The delay time is proportional to the capacitor value.

P_{RISE} Power-Up Rise Time Control. A capacitor connected between this pin and ground controls the rate at which the regulator output rises after the regulator has been enabled. The rise time is proportional to the value of the capacitor.

PE Power Up Enable. This open-collector output goes high when the regulator output voltage has reached its full value. It can be used as a signal to the user that the regulator is operating or to enable other devices.

The AM6300 can be used as a 723 type voltage regulator (refer to Fig. 6-21) with the following equation establishing the output voltage:

$$\frac{V_{OUT}}{V_{REF}} = \frac{R1 + R2}{R2} \qquad \textbf{Equation 6-1}$$

where $V_{REF} = 2.5$ V

Current sensing with resistor R_s, as shown in Fig. 6-21, puts the resistor outside of the feedback loop between the IS^+ and IS^- terminals. This reduces power dissipation of the chip. R_s is equal to $V_{cl}/2\ I_{cl}$. The current-limit voltage, $V_{Cl/2}$, is the maximum voltage allowed across the current-sense amplifier, refer to Fig. 6-20. This is approximately 0.2 V so a voltage drop across R_s of 0.2 V or greater triggers the internal comparator and shuts down the power supply.

The amount of time before shutdown is also programmable. With no capacitor attached to the I_{DEL} terminal to ground, it is 1.5 microseconds. With a capacitor, it is about 0.22 microseconds per picofarad plus the 1.5 microsecond delay already inherently there. When the AM6300 is shut down by overcurrent, its I_{ALARM} flag goes high (pin 12).

The under and overvoltage sensors also shut down the AM6300 when the output voltage goes beyond the programmed limits. If the O_{SEL} has a logic high on it, a 5 percent limit is in effect. A logic low results in a 10 percent limit. With a capacitor on the V_{DEL} pin, a delay is experienced before device shutdown. The V_{DEL} capacitor value and inherent built-in delay is identical to the already mentioned I_{DEL} pin. After shutdown, either by overvoltage or current, the AM6300 can be reset by two methods. First, power can be momentarily removed and reapplied or the $\overline{P}_{DN}$ or $\overline{P}_{DR}$ pins could have a low followed by a high placed on them. A reset switch on these pins driving them to ground would also work.

A user-defined delay is possible with respect to the time required for resetting the AM6300 by connecting a capacitor from $\overline{P}_{DEL}$ to ground and taking $\overline{P}_{DR}$ low and then high. As was pointed out in the description of the switching voltage regulator ICs, the AM6300 also has a soft-start feature with linear rate of change by attaching a capacitor from the P_{RISE} terminal to ground.

The AM6300 can be used in switching voltage regulator applications with a self-oscillating switch (Fig. 6-22) and requires no pulse generator. The output voltage is defined as

$$V_{OUT} = V_{in} (T_1/T) \qquad \textbf{Equation 6-2}$$

where T_1 = time the device is on
T = total time period

The power enable (PE) function should not be used in this type of application.

As discussed previously, many systems require power-on sequencing schemes. The circuit in Fig. 6-23 is a sequencing priority circuit in which the PE (power enable) pin of one AM6300 is connected to a digital input of another AM6300. The second AM6300 is powered up only when the first AM6300 is switched on and its output voltage reaches its final value.

Another method of accomplishing this would be to connect

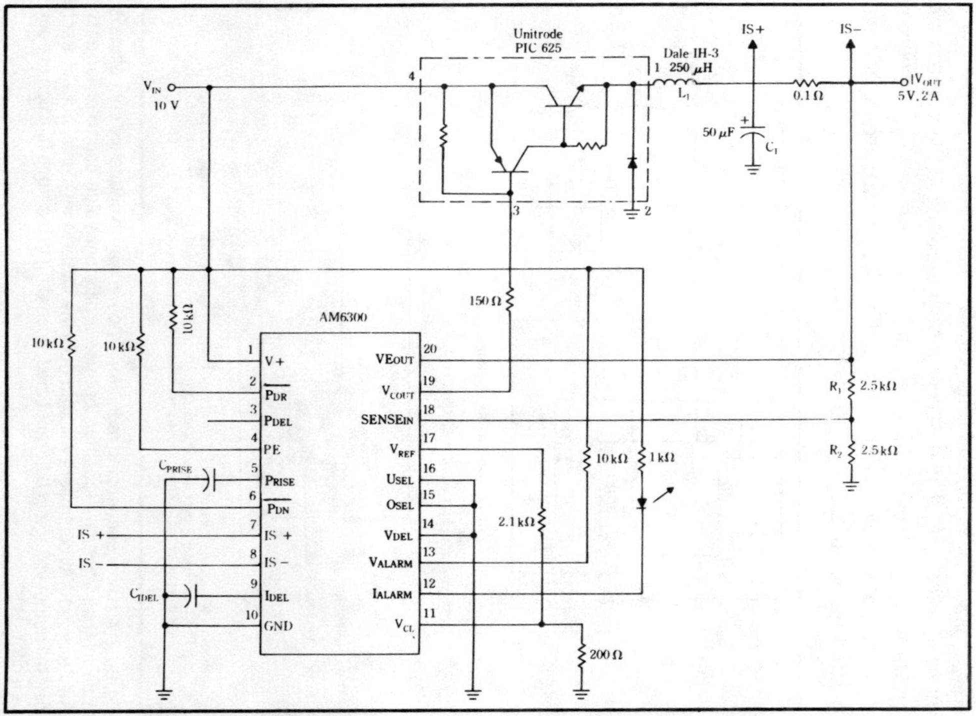

Fig. 6-22. A switching regulator AM6300-based power supply.

different value capacitors to PDEL. When the regulator with the larger capacitor is powered up it will come up later than the regulator with the smaller capacitor. Also, a power-down sequence could be accomplished in which the regulators power down in an order different from the order in which they powered up by using the input (either $\overline{P}_{DN}$ or $\overline{P}_{DR}$) that was not used for powering up.

The MC3424

The MC3424 over/undervoltage-protection circuit is a 16-pin DIP IC manufactured by Motorola. This IC (Fig. 6-24) is a high-current, dual-channel, voltage comparator capable of sensing positive or negative voltages either greater than or less than the 2.5 V internal reference. The device can be programmed to operate with or without hysteresis. Each channel has a current-limited output driver capable of sourcing 300 mA. The outputs are protected from turning on when subjected to power supply transients.

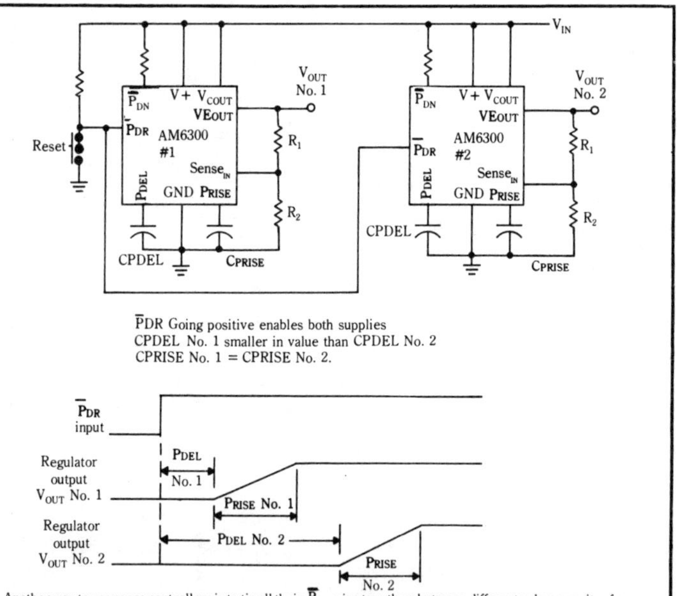

Another way to sequence controllers is to tie all their $\overline{P}_{DR}$ pins together, but use a different value capacitor for each PDEL terminal. All controllers will start from the same reset signal, but will reach their full output voltage at different times, according to the ramp function of each capacitor.

Fig. 6-23. A power sequencing regulator. Another way to sequence controllers is to tie all their $\overline{P}_{DR}$ pins together, but use a different value capacitor for each PDEL terminal. All controllers will start from the same reset signal, but will reach their full output voltage at different times, according to the ramp function of each capacitor.

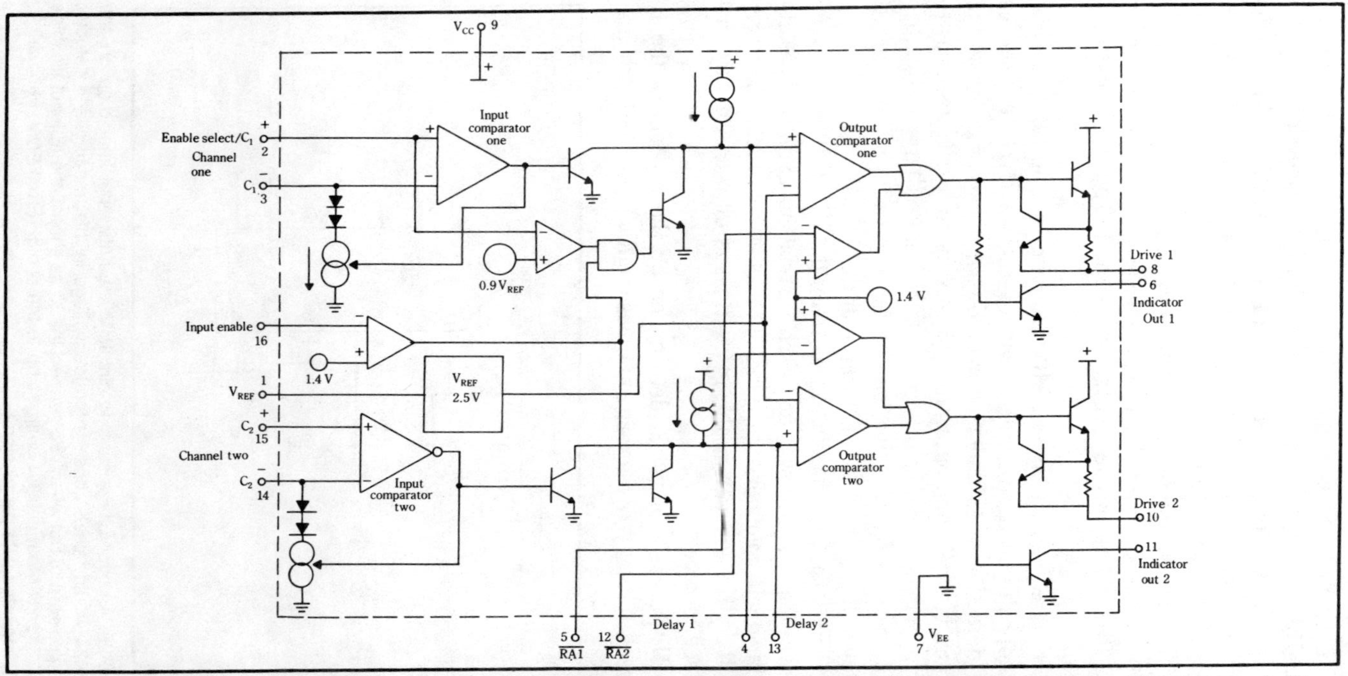

Fig. 6-24. The MC3424 block diagram. Sensing over/undervoltages on two channels, the MC3424/3524 operates with or without hysteresis. (From *EDN*, January 7, 1981; reprinted with permission.)

Each channel has an individual indicator capable of sinking 30 mA. This allows remote activation which can override the main inputs (the C inputs) at any time to turn on the power outputs (DRIVE 1 and 2). There is also an Input Enable pin which disables one or both comparators.

The delay time before tripping the outputs when an under or overvoltage occurs is also programmable. This delay is variable from 1 μs to 1 ms by use of capacitors in the 100 picofarad to 0.1 μF range.

Figure 6-25 shows the MC3424 with its number two comparator half sensing overvoltage with hysteresis. Hysteresis means that detection of an under or overvoltage does not occur at one distinct voltage point but rather that the monitored voltage has to pass through a *band of voltages* usually centered around this trip point, before circuit activation occurs. The width of this band of voltages (the hysteresis) is programmable.

The R1 C1 product in Fig. 6-25 on the IE (Input Enable) pin ensures that during power-up (when the final input voltage is establishing itself) an undervoltage indication does not occur. The larger the R1 C1 combination, the more time is allowed for the final voltage to occur.

Hysteresis voltage is determined by the product of the I_H current (typically 12 microamps) times the R_A/R_B parallel combina-

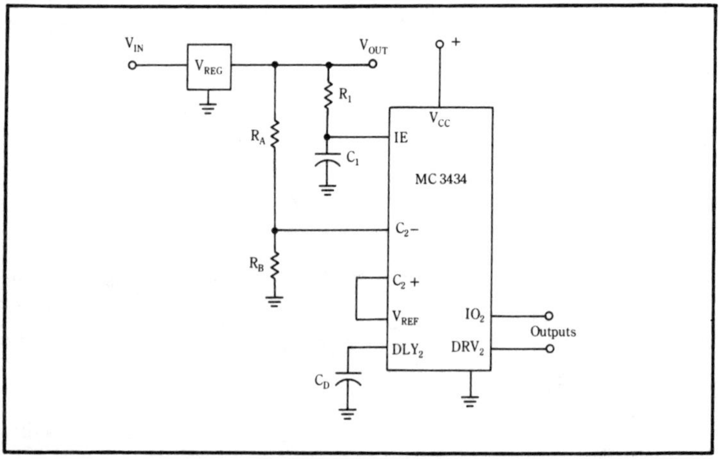

Fig. 6-25. An overvoltage sensing circuit. Without its input-enable option, the MC3424 would falsely sense a too-low voltage in this undervoltage detector during power-up. But when you connect the IC as shown, the IE input disables the C^- sense input until C1 charges up to threshold. (From *EDN*, February 4, 1981; reprinted with permission.)

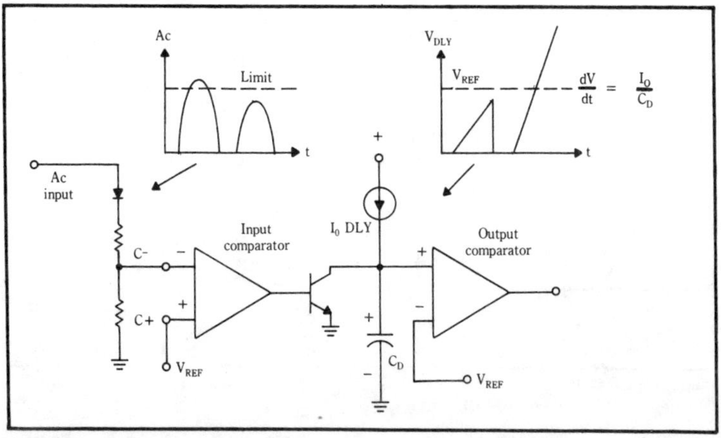

Fig. 6-26. Detecting ac voltage levels. Ac-power-line monitoring can occur on a cycle-by-cycle basis with the MC3424. If the peak ac input doesn't exceed a predetermined limit, capacitor C_D charges to the output comparator's reference level, and a fault output results. (From *EDN*, February 4, 1981; reprinted with permission.)

tion. The internal charge current, I_0, which is approximately 200 microamps, and capacitor C_D control the comparator output delay time.

Ac can be monitored by the MC3424 as well. Figure 6-26 shows how low ac occurring for any specified number of cycles can be detected. When the ac voltage is below the defined limit, the device's input comparator allows the delay capacitor to charge at a dv/dt (rate of voltage per unit of time) determined by I_0/C_D.

If the following cycle's most positive point exceeds the limit, the comparator's output tends to pull the C_D capacitor's charge back to ground. If the limit is not reached, the charging of C_D continues to ramp up at an uninterrupted rate. The selection of the value of C_D determines how many cycles can be missed before this capacitor voltage reaches the threshold voltage of the output comparator's V_{REF} and triggers an output response.

Just as with the AM6300, the MC3424 can be used in multiple power monitoring or other applications with sequential turn-on. The MC3424 could have one channel's I/O pin connected to the other's RA or DLY (delay) terminal. As an example, if channel one's I/O pin is connected to channel two's output, two won't activate one's output. However, if channel one's I/O is connected to channel two's DLY input so that one's output holds DLY low, a channel one output inhibits a channel two output without the converse condition being possible.

Chapter 7

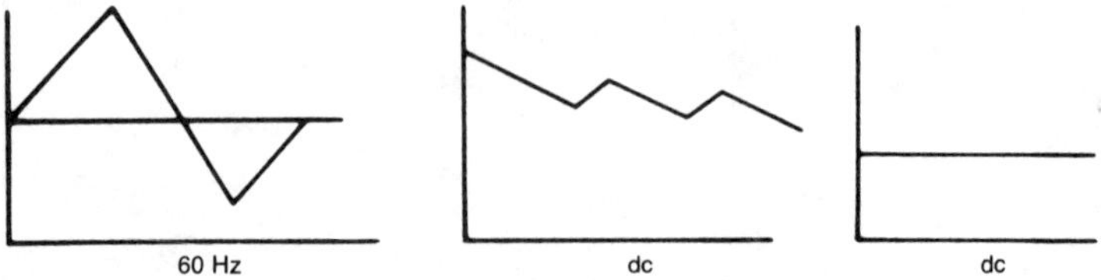

Power Conserving Techniques

A Frost and Sullivan research study in 1975 indicated that for every dollar spent to conserve energy, there would be a two dollar savings. Industry began realizing this and the energy conservation business mushroomed to become a billion dollar business. Table 7-1 is a list of sales by product in this industry.

Thus far, we have examined numerous voltage regulator integrated circuits and applications using them. Let us now focus on using them in the most power-efficient manner possible. Most of this concern revolves around battery-powered equipment because of the emergence of electronic monitoring equipment at third world sites where electricity does not exist. Also, there are now numerous solar battery charging schemes in which power conservation is at a premium. Examples of this are solar-powered calculators and an all-electronic solar powered camera by Ricoh.

This chapter examines eight areas in which efforts can be made to conserve power while the product is still in the design phase. These are as follows:

- Display blanking and power strobing of light-emitting diodes and other similar optoelectronic devices.
- Novel uses of the LM3909 Power-Miser flasher IC.
- Judicious memory management schemes.
- Intentional shutdown of microprocessors.
- Examination of CMOS as a power-saving logic family.
- Minimizing voltage regulator dropout voltages.
- Examination of on-demand power supplies.
- Generation of multiple and negative voltages from a positive

Table 7-1. Energy Conservation Sales by Product Listing. (From *EDN*, June 5, 1975; reprinted with permission.)

Product Category	1974	1975	1976	1978	1980	1985
Electrical Energy Controls	40	47	55	70	90	125
Power-Factor Capacitors	(8)			(14)		
Power-Demand Controls	(22)			(42)		
Other	(10)			(14)		
Heat Energy Controls	55	62	72	95	115	160
Combustion Controls	(41)			(72)		
Heat-Recovery Controls	(14)			(23)		
Specialized Energy Controls	5	6	8	10	13	20
Computers in Energy Control	30	40	50	70	90	150
Mini-Computers	(25)			(44)		
Micro-Computers	(1)			(20)		
Other	(4)			(6)		
Total	130	155	184	245	308	455

* Millions of Dollars

unipolar supply without adding transformer taps or employing switching regulator techniques.

Power will be saved by employing one or more of these eight techniques. This has an impact on other product features. The transformer can be smaller, the diodes less hefty, the overall weight less, and the cabinet smaller. In addition, less heatsinking will be required. The list could be extended further, but the real bottom line is cost reduction as a result of each of these features. This also helps the manufacturer make the product more cost competitive!

DISPLAY BLANKING AND RELATED OPTOELECTRONIC TECHNIQUES

An LED (light-emitting diode) can be supplied with pulses instead of a continuous steady supply of dc voltage. The result is the ability to allow higher current pulses of less sustained duration (lower duty cycle) to drive an LED with apparent greater brightness. This phenomenon, known as *light enhancement,* is partly due to the eye's retention of high brightness levels.

Multiplexing of 7-segment LEDs is very common. This means that a circuit, as shown in Fig. 7-1, will have only one transistor turned on at a time. A square wave with a 25 percent duty cycle will be alternately applied to each transistor base in a nonoverlapping manner. As each transistor turns on, it conducts, allowing the current through each of the conducting LED segments to flow through each transistor to ground. If your eye were fast enough, or if the scanning rate of the square wave pulses were slow enough, you could see this. The resulting power savings is simple—if the four displays are only on one-fourth of the time, despite appearing

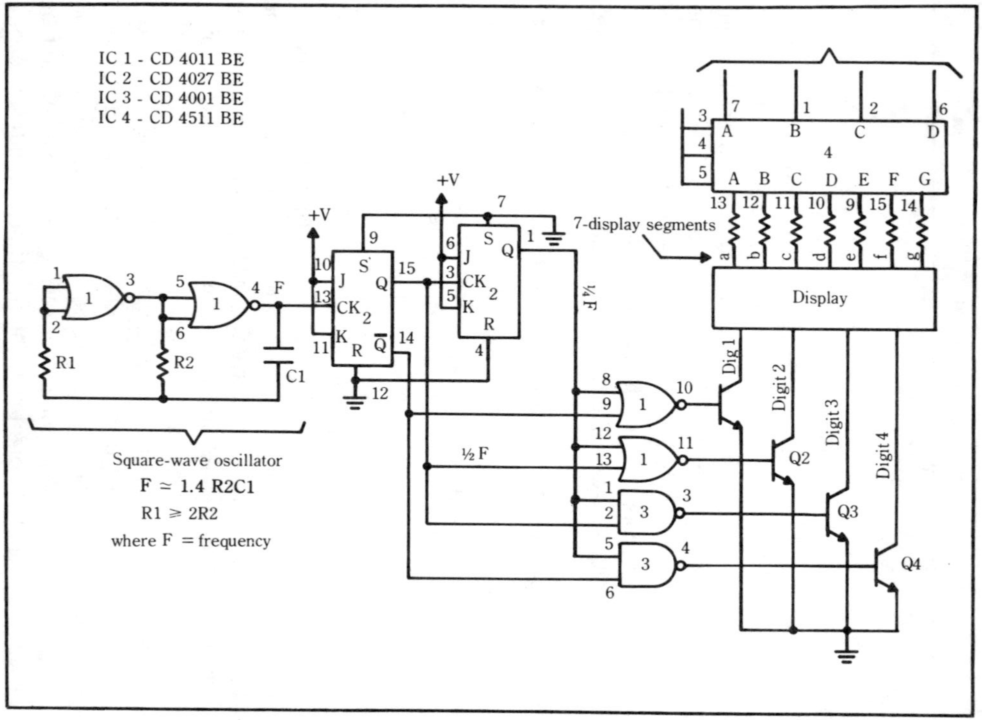

Fig. 7-1. Multiplexing seven-segment displays.

otherwise then a like amount of only 25 percent of the previously required power is consumed.

Digital voltmeters (DVMs) with 3, 3½, and 4½ digit LED readouts can consume a lot of power, even with their displays being power-strobed or multiplexed. One way to further curb this power consumption is to suppress the initial zeros. Until the D/A converter manufacturers become energy conscious by blanking their initial zeros with these built-in functions on the chip, an external solution will have to be sought. The following two circuits show how to do this.

The Motorola MC14433 A/D converter is designed for DVM applications. This converter scans the display from the most to the least significant digit, that is, from left to right. The circuit in Fig. 7-2 detects when an initial digit is a zero and clocks IC1A to a logic high which the CD4011 two-input NAND gate inverts. This signal is the decoder blanking input. If the second and third digits are zeros, the display is left in the blanked state. If any one of the digit codes is not a zero, IC1A resets and blanking is inhibited.

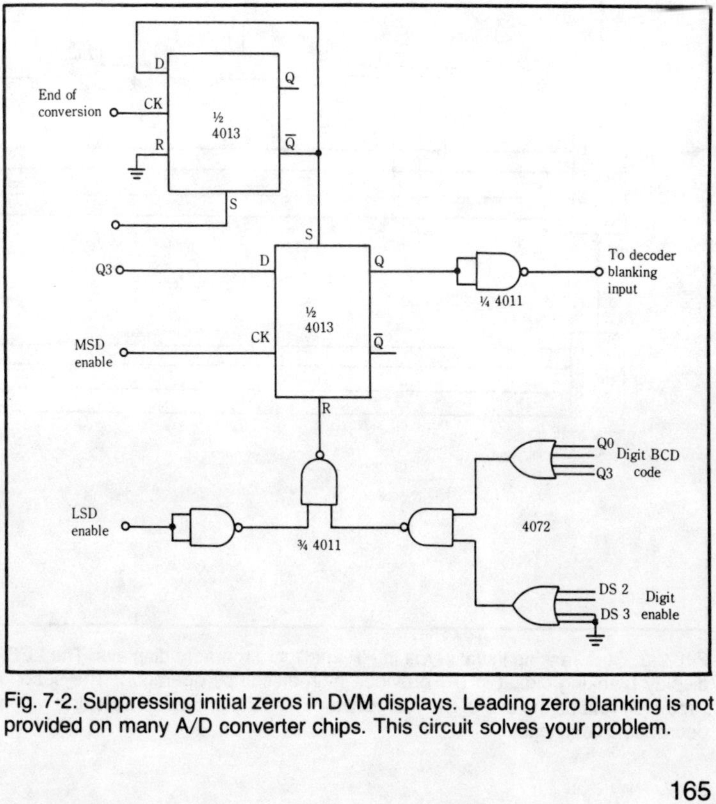

Fig. 7-2. Suppressing initial zeros in DVM displays. Leading zero blanking is not provided on many A/D converter chips. This circuit solves your problem.

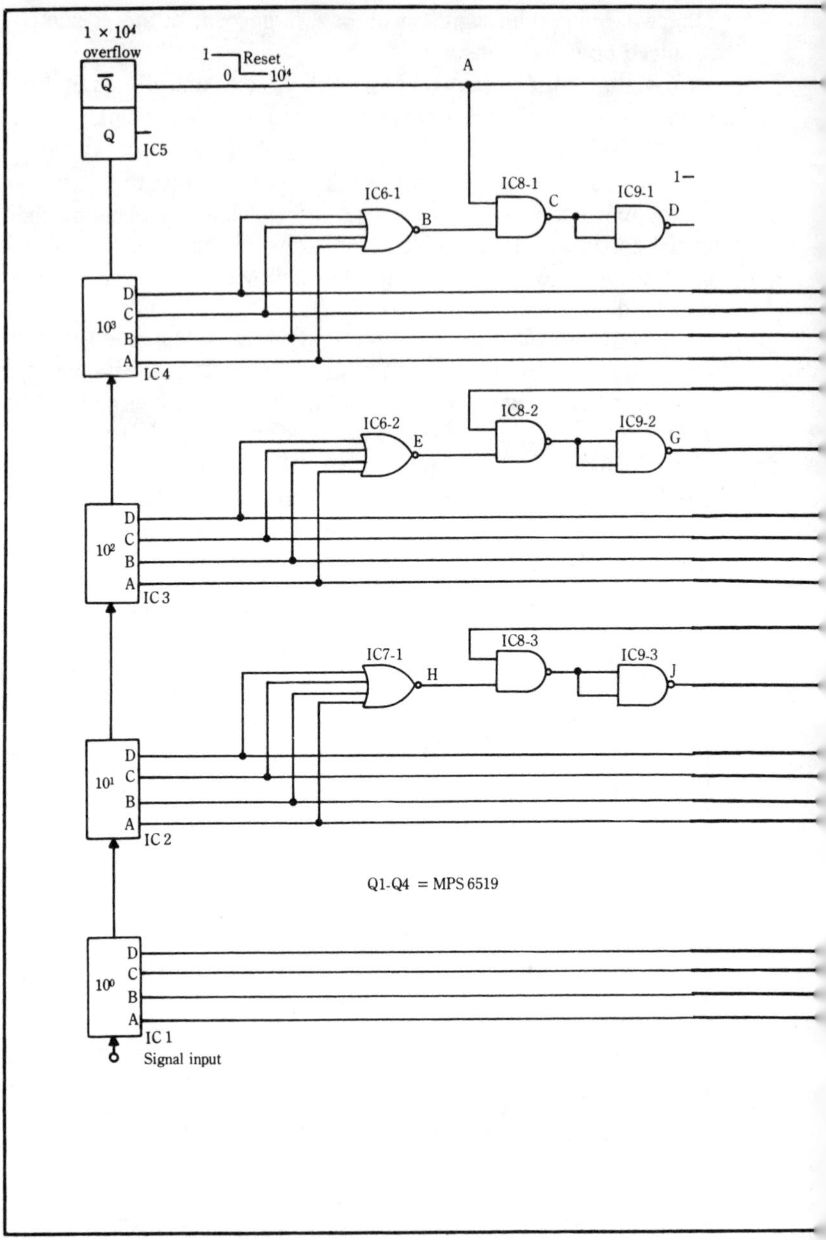

Fig. 7-3. Suppressing initial zeros in HP intelligent memory displays. The LED-display blanking circuit in (A) provides high-efficiency operation. The second circuit (B) retains the use of the memory function and allows multiplexing with a decrease in efficiency.

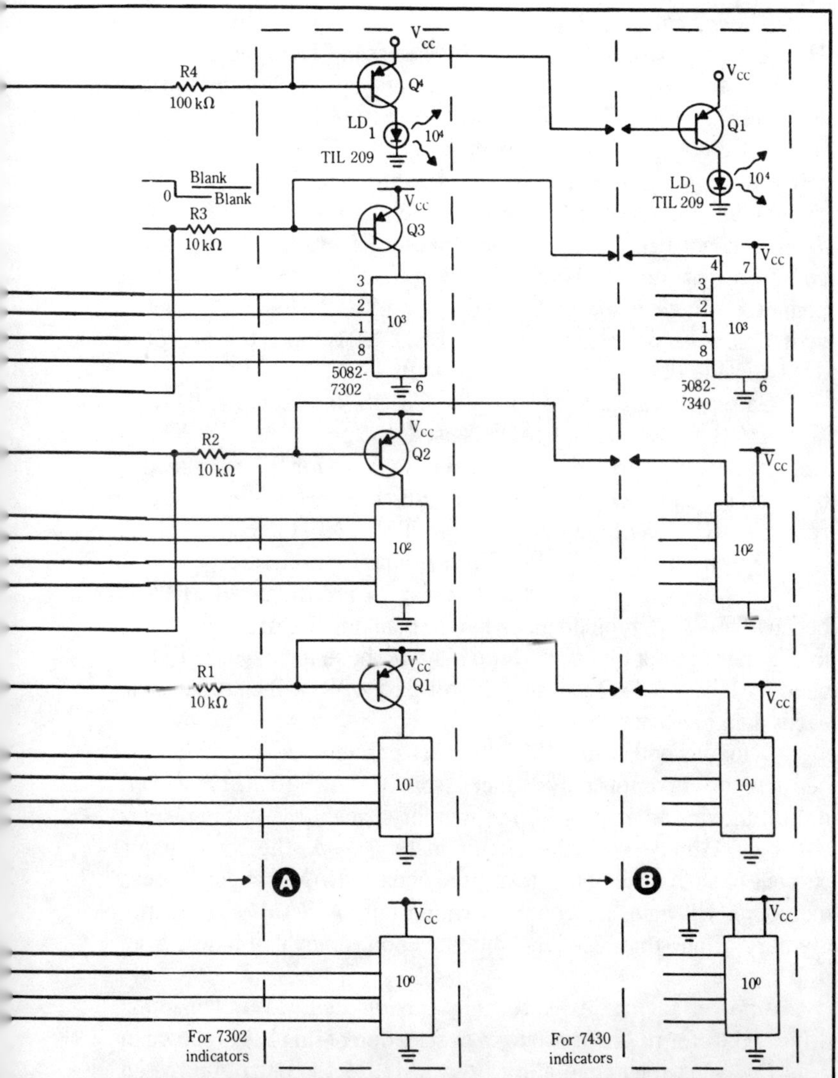

Stored count	Points in the circuit										Display condition				
	A	B	C	D	E	F	G	H	I	J	10^0	10^1	10^2	10^3	10^4
00000	1	1	0	1	1	0	1	1	0	1	√	X	X	X	X
10000	0	1	1	0	1	1	0	1	1	0	√	√	√	√	√
01010	1	0	1	0	1	1	0	0	1	0	√	√	√	√	X
00001	1	1	0	1	1	0	1	1	0	1	√	X	X	X	X

(Note: √ = display enabled; X = display inhibited)

The circuits in Fig. 7-3 do a similar job using only the Hewlett-Packard 4½ digit 5082-7300 series of LED indicators. This series of 4 × 7 dot matrix indicators can display numbers up to 15, or F in hexadecimal. They have all the required drive/decode and memory circuitry built into the display. The 7300/7302 types have left and right-hand decimal points while the 7340 type has a display-blanking feature instead of decimal points. Refer to Fig. 7-3A and note ICs 1-4. These represent four decades of ripple counters that supply the BCD inputs. Transistors Q1 to Q3 are connected in series with V_{cc} and pin 7 of each device and act as switches. A conventional two-lead LED, LD1, indicates overflow. This LED is driven by either a D or JK latch.

Gates within ICs 6 to 8 may be CMOS, for example, CD4002 and CD4011; however, if CMOS is used, IC9 should be a Darlington transistor pair to provide additional base drive for their respective transistors. This added drive is a result of the 94 mA each display requires. IC9 may be a low powered TTL SN74L00N.

The circuit in Fig. 7-3B is quite similar to its counterpart, with the exception that no driving transistors are required. This is because the 7340 type displays have a blanking feature. The blanking signals are applied directly to pin 4 of these displays. If TTL is used for IC9, a 3.3 kΩ pull-up resistor tied to V_{cc} will be required at each blanking input.

Blanking with the 7430 is not as efficient as with the 7302 because the current is only reduced from 94 mA to 45 mA; however, if multiplexing is employed, the memory capability of this display will be retained. With the circuit in Fig. 7-3A, the total supply current of each display is interrupted and the RAM (random access memory) will have its contents wiped out. A RAM is a volatile memory or one that loses its contents upon removal of power from it.

A power-saving design technique using the LM3914 bar-graph driver IC (refer to Chapter 6 for a description of this IC) is shown in Fig. 7-4. This technique allows the LM3914 to control up to ten LEDs in series with just 10 mA. The LM3914 draws 5 mA itself. Normally ten LEDs drawing 10 mA each would draw 100 mA of current, but by placing the LEDs in series, a substantial power savings is realized. The high-efficiency red LEDs, as well as yellow and green LEDs, have an approximate 0.3 V additional drop across them, so a +24 V regulated source may be required instead of the +22 V supply as shown.

The circuit in Fig. 7-5 uses an ICL7107 3½ digit A/D converter

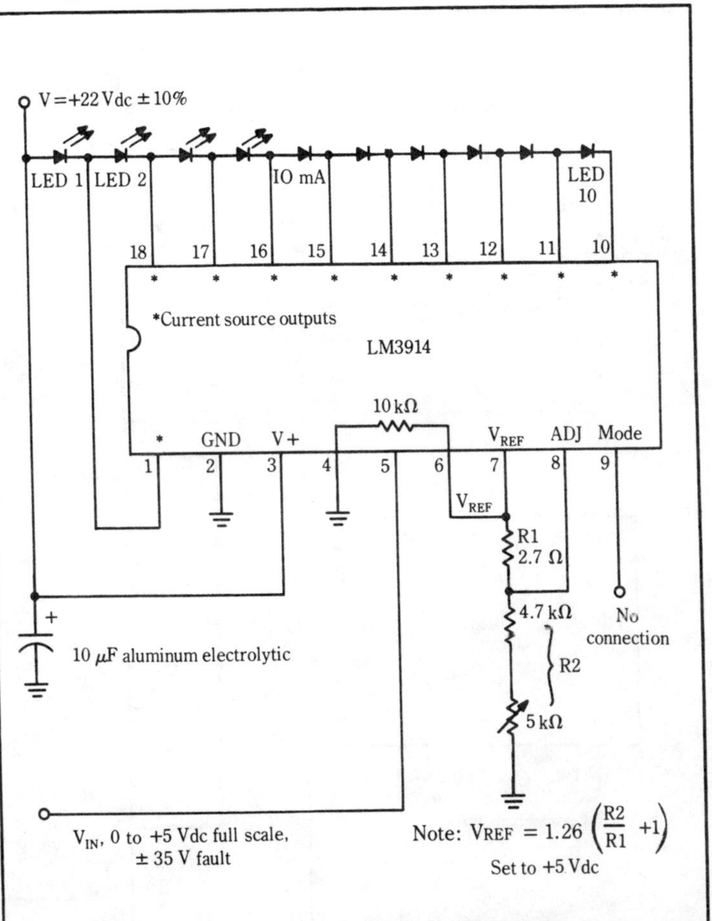

Fig. 7-4. Saving power with series LEDs in an LM3914 application. The ten current sources within the LM3914 bar-graph chip are individually turned on and off by the amplitude of V_{IN}. Since only one current source may be on at a time in the "dot" mode, 10 mA may be routed through any series combination of up to ten LEDs. Here, five LEDs are on, indicating that V_{IN} is between 2.5 and 3.0 V.

IC with built-in display drivers (19 lines), each with an 8 mA constant-current sink to drive LED displays. If we realistically assume a 2.0 V drop across each LED segment, that leaves 3.0 V left from the +5.0 V supply to be dropped across the output transistor. The insertion of the three diodes from pin 1 to the LEDs allows for three 0.6 V drops to occur and reduces the output power dissipation from a worse case of 600 mW to only 120 mW while providing a brightness control selector switch.

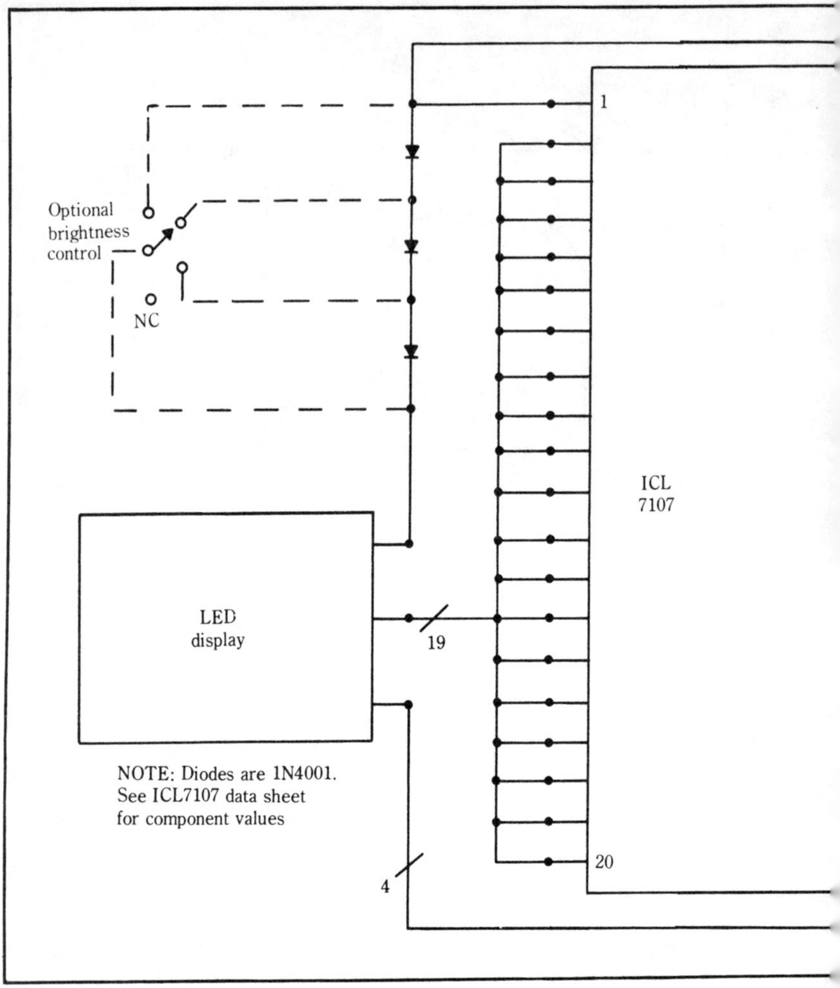

Fig. 7-5. Saving power in an ICL7107 3½ digit LED driver circuit (From *EDN*, April 5, 1980; reprinted with permission.)

LM3909 POWER-MISER IC

This IC is specifically designed to flash LEDs with a supply voltage of 1.5 V or less (one battery). Using timing capacitors for voltage boosting enables generation of 2 V or greater pulses. The IC is self-starting so it requires only an LED, capacitor and battery. The power-miserly nature of the IC is shown in Table 7-2 for the circuit shown in Fig. 7-6 which flashes at 1 Hz. Figure 7-7 shows a power failure alarm that will last for three years on a single "D"

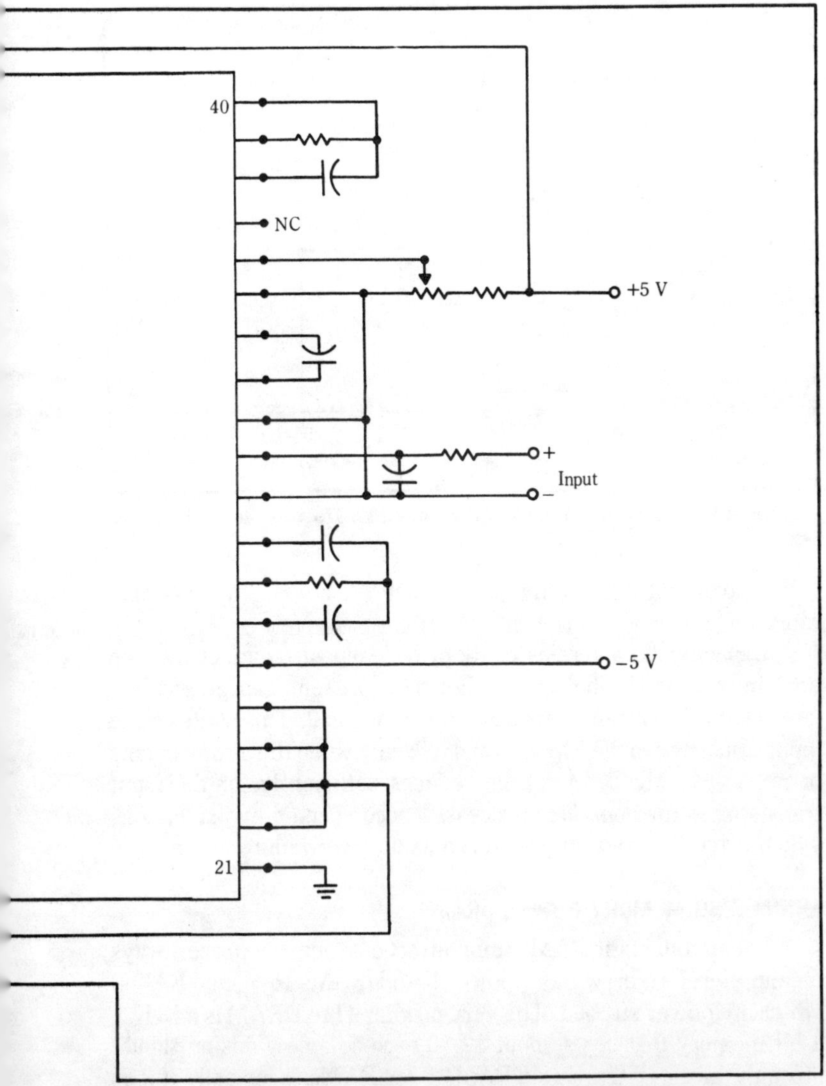

Table 7-2. The LM3909 Power Consumption vs. Battery Life.

SIZE CELL	TYPE	
	Standard	Alkaline
AA	3 months	6 months
C	7 months	15 months
D	1.3 years	2.6 years

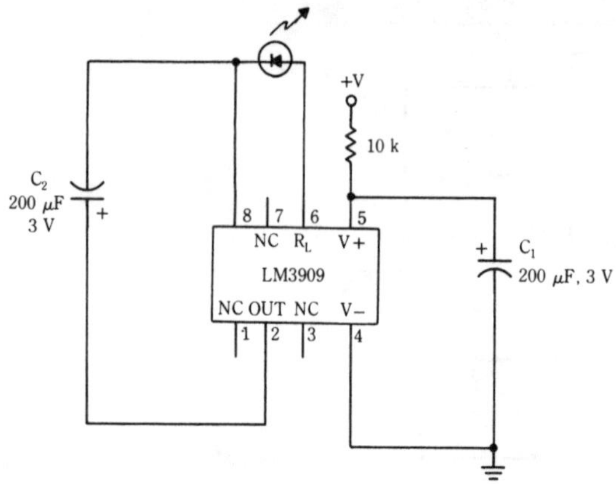

Providing ON indication for any circuit that uses 4 to 40 V, this flasher draws only about 0.6 mA.

Fig. 7-6. A 1.5 V LED flasher. (From *EDN*, April 20, 1979; reprinted with permission.)

cell. The circuit draws only 50 μA when the ac is within a normal range.

Capacitor C1 and resistor R2 provide a positive feedback loop enabling oscillation; however, when ac is present, charge accumulates on pin 8 to make it more positive than pin 2. This shuts off the input transistor to the LM3909 and it is only when the ac power fails or goes away that C1s charge reduces sufficiently for the input transistor to turn on. This process is accompanied by oscillations which drive the speaker and serve as a power failure alarm.

MINIMIZING MEMORY POWER DRAIN

In general, static RAMs (random access memories) are always on and therefore draw more power than DRAMs (dynamic RAMs) which are power strobed. The very popular 4116 DRAM is a 16 K × 1 bit memory that costs about $2.50 each so a new design should strongly investigate using DRAMs over RAMs, especially if it is battery powered or if power consumption is a concern.

PROMs (programmable read only memories) that are bipolar often are avoided in battery-operated designs because of their power consumption. The circuit in Fig. 7-8 uses a bipolar PROM with a power-saving design technique to decode a pocket calculator's 7-segment output into BCD code. Segment outputs SA through SG are ORed together into signal SS which is a pulse train with a 1:1 mark/space ratio and period of 350 microseconds.

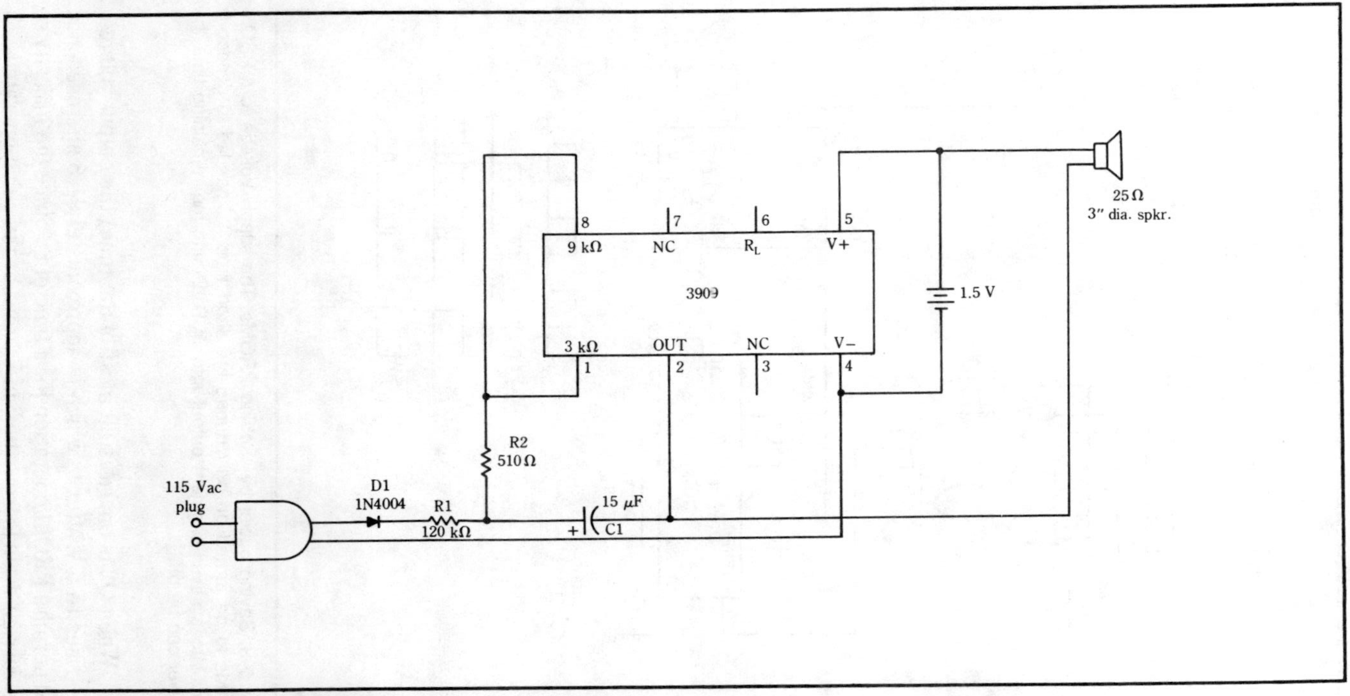

Fig. 7-7. An LM3909 low-drain power failure alarm.

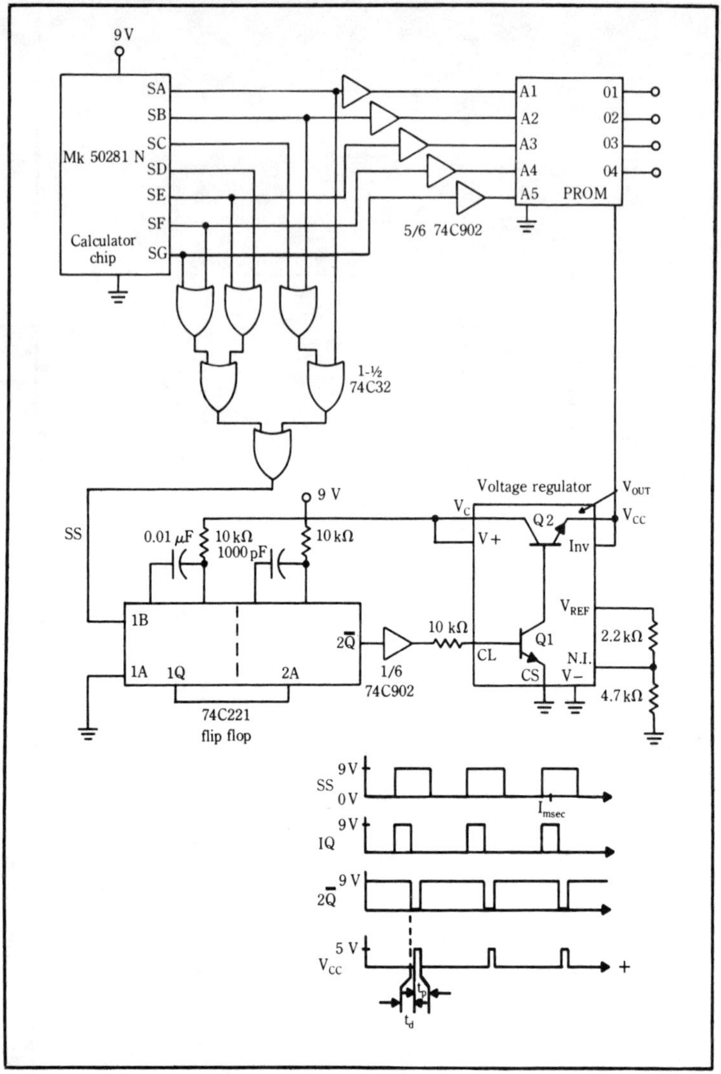

Fig. 7-8. Saving power by strobing PROMs. The supply voltage to a PROM needs to be turned on only during the short time, t_p, needed to decode a calculator's seven-segment output into BCD. During the remaining time the power can be off.

When Q2 is on, (it is off most of the time) the output voltage, V_{cc}, rises to 5 V after t_d which is approximately 1.5 microseconds and permits PROM operation for a time of t_p. Since the majority of the current drain is from the 3 mA drawn by the pulse-operated 723 voltage regulator, the 35 mA PROM current normally experienced with continuous operation is reduced by 90 percent.

MICROPROCESSOR SHUTDOWN

After a control sequence, it is often advantageous to shut the microprocessor down to save power. The circuit in Fig. 7-9 does this.

Ensuring that the microprocessor is off also guarantees against voltages drifting down to 2 V or 3 V, where logic states are erroneous and a turn-on command could result in turning on equipment that was not intended to come on.

The cross-coupled NAND gate latch activates the processor when the START input is pulled to ground.

This places a logic high on the latch's Q output and drives the zener-regulated pass transistor into conduction. This allows the processor's supply voltage to rise to its normal value.

When the processor generates a command to shutdown, the logic low on the $\overline{\text{KILL}}$ line pulls the base of the series-pass transistor toward ground. This turns the power supply off. Once the supply is turned off, any erratic signals on the microprocessor's output lines will have no effect on it.

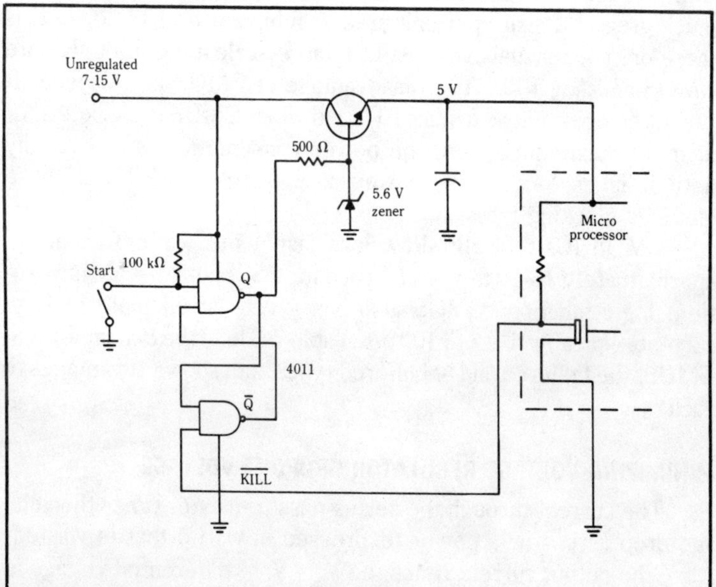

Fig. 7-9. A microprocessor shutdown circuit.

Table 7-3. Unbuffered vs Buffered CMOS Characteristics.

Characteristic	Buffered Version ("B")	Unbuffered Version ("UB")
Propagation Delay (Speed)	Moderate	Fast
Noise Immunity/Margin	Excellent	Good
Output Impedance and Output Transition Time	Constant	Variable
Ac Gain	High	Low
Output Oscillation for Slow Inputs	Yes	No
Input Capacitance	Low	High

THE POWER STINGY CMOS FAMILY

CMOS is a very versatile logic family with numerous advantages over the earlier TTL and DTL families. The most notable advantages are low-power consumption, very large fan-out (typically one CMOS output can drive 50 CMOS inputs), excellent noise immunity and operation over a variable voltage range of 3 Vdc to 20 Vdc. TTL (transistor-transistor logic) operates only off of +5 V. The process by which CMOS is made also enables it to have very much greater density per unit area than bipolar or TTL logic. It is therefore not unusual that most LSI (large-scale integration) ICs are CMOS and not TTL. The disadvantage of CMOS is its speed. It cannot be operated as fast as TTL and when CMOS is clocked at or near its maximum speed, the power consumption grows rapidly until it is no longer less power consuming than TTL at these elevated clocking rates.

CMOS ICs typically draw less than 1 mA while TTL draws approximately 16-20 mA per IC pack for a SSI (small-scale integration) logic function. CMOS also has a very small input leakage current, typically 10 pA to 100 pA. Table 7-3 lists the two families of CMOS, the buffered and unbuffered types, and shows advantages of each.

MINIMIZING VOLTAGE REGULATOR DROPOUT VOLTAGE

The current through the series-pass transistor times the voltage drop across it is power (expressed in watts) that is wasted. Also, the output current times the $V_{IN} - V_{OUT}$ differential voltage is the minimum voltage difference in a voltage regulator between its input and output whereby regulation can still be maintained. Operation at less than this differential results in poor regulation or else the voltage regulator "drops out" of regulation. Therefore, this voltage is often referred to as "dropout" voltage. A typical voltage regulator like the LM317 has a three volt or greater dropout voltage.

A transistor has a V_{BE} (SAT) typically of less than 1 V, at higher currents V_{BE} (SAT) might approach 2 V—depending on the type of transistor. In the effort to conserve power, the voltage regulator's dropout voltage as well as the V_{BE} (SAT) of transistor has to be minimized.

The circuit in Fig. 7-10 is a discrete voltage regulator design with a dropout voltage of less than 1 V. The 741 op amp is a voltage-to-current amplifier. Its output is current-amplified or buffered to drive the load by the two transistors. The feedback provided by the R3/R4 potential divider samples this voltage with the reference voltage generated by zener diode, Z1.

The zener diode, Z1, is biased by R1, and the R2-C1 combination provides low-pass filtering for any noise or ripple. Resistor R5 limits the output current from the op amp and can be omitted if you are not concerned about excessive current demands by the load. The output voltage of this efficient discrete regulator can be adjusted from 3 V to 10 V by the (R3+R4):(R3) ratio or by connecting a pot across the zener and applying the wiper of the pot to pin 3 of the op amp.

The circuit in Fig. 7-11 is specifically designed for lithium batteries to accommodate their low terminal voltage at lower temperatures. This circuit provides a stable +5.0 V output at 10 mA

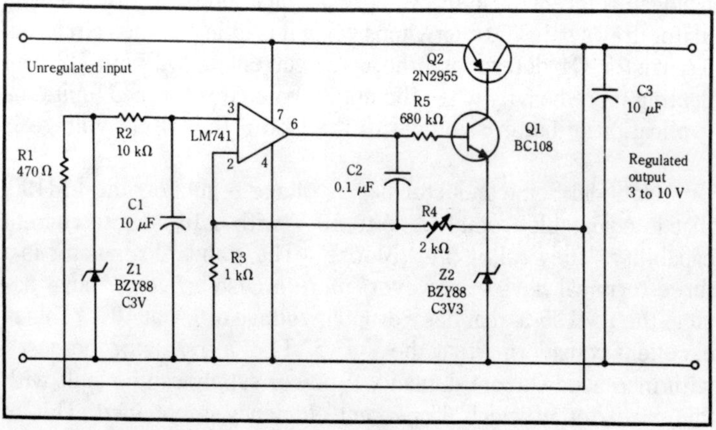

Fig. 7-10. Reducing voltage regulator dropout.

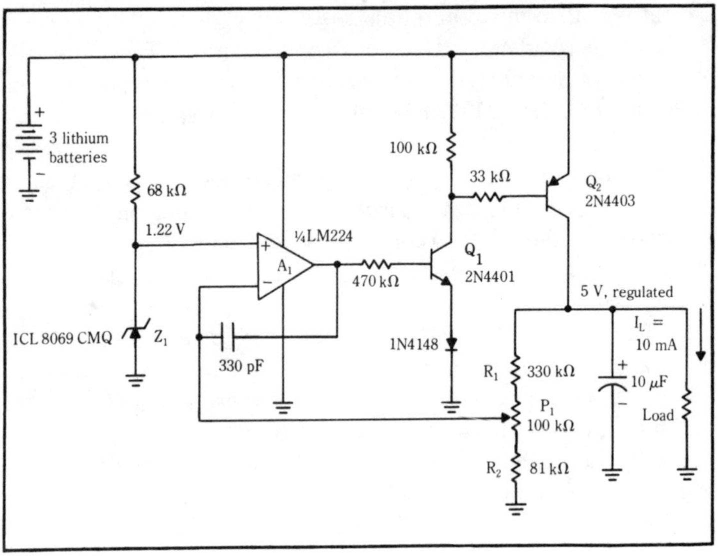

Fig. 7-11. Reducing voltage regulator dropout in a lithium battery circuit. (From *Electronics*, April 12, 1979; reprinted with permission.)

with a supply voltage varying from 10 V down to as low as +5.2 V. This low dropout voltage (0.2 V) is obtained by operating the output transistor Q1, in the common-emitter mode of operation. The resulting collector-to-emitter voltage drop is much lower than the base-to-emitter voltage drop (typically 0.65 V) of transistors in the emitter follower mode of operation, as is common practice for most voltage regulators. The low-voltage "zener" diode is actually a bandgap reference and draws only 50 microamps. The whole regulator draws only 250 microamps while it is idling or open circuited. Transistor Q1 determines the drive current to Q2 with Q2 conducting more heavily when the output voltage is low and limits the application of battery voltage to the load if the output voltage is high.

National Semiconductor has a voltage regulator, the LM196, that is adjustable from 1.25 V to 15 V with a 10 ampere current capability. They call it the "Moose". The standard regulator is a three-terminal device; however, there is also a five-terminal device, the LM196-5, that has a dropout voltage of 1 V at 10 A. This is excellent considering that the V_{BE} (SAT) of a transistor increases with increased current demand. A power supply can be built with this regulator in which 65 percent efficiency is obtained. This is unprecedented for IC linear voltage regulators.

ON DEMAND POWER SUPPLY

Ideally, a circuit should never come on until it is needed. The very simple circuit in Fig. 7-12 charges storage capacitor C1 on each positive half cycle, until the charge voltage equals the zener reference voltage. At this level, the gate of the SCR is reverse biased and is turned off at the end of the cycle. If the load begins to draw current, the voltage on the storage capacitor falls below the zener voltage and the gate of the SCR is forward biased and turns on power to the load (the "demand" function). This circuit's relaxation mode of oscillation regulation places a low-frequency ripple on the output. This ripple can be reduced by making the R2C1 time constant fairly large. R2 should be large anyway to minimize power drain.

GENERATING MULTIPLE AND NEGATIVE VOLTAGES FROM A SINGLE SUPPLY

Often there is a requirement for −12 V or −5 V at some very small current, typically 2 mA or less. What do you do? Adding an additional negative regulator, transformer winding, and filter capacitor seems a bother, especially if you only have +5 V on the PC card. The answer lies in the use of some very efficient (up to 99 percent), versatile, voltage regulators, the ICL7660/3/4/5 series made by Intersil. The ICL7660 converts +5 V to ±5 V or does voltage multiplication as a function of the ratio of two capacitors. Refer to the ICL7660 block diagram in Fig. 7-13 and note how during the two phases of operation alternate pairs of switches are

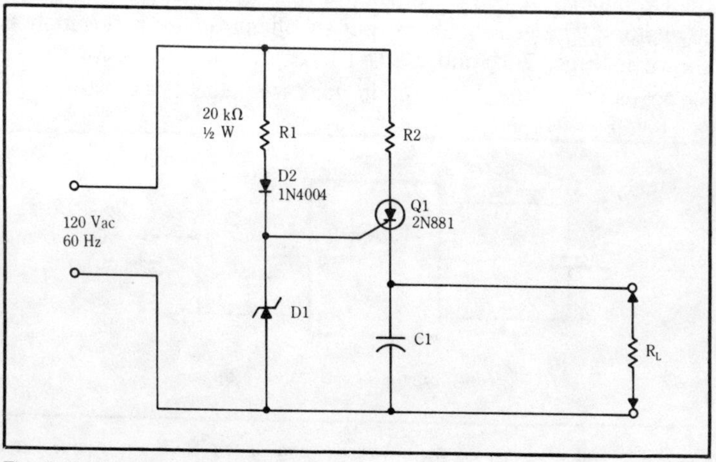

Fig. 7-12. A power on demand circuit.

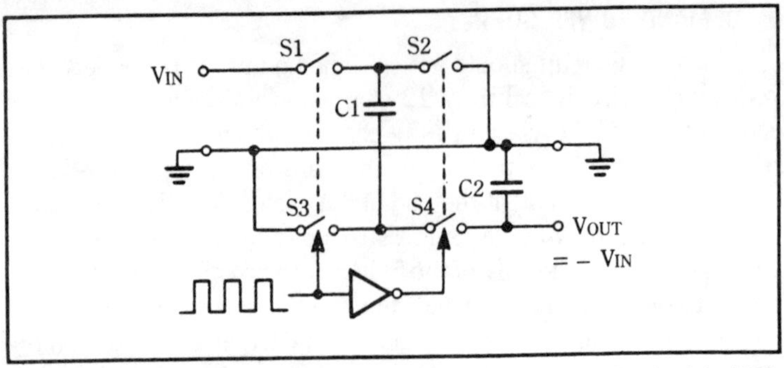

Fig. 7-13. The ICL7660 block diagram. (From Intersil Data Book, July 1979; reprinted with permission.)

closed, transferring charge, as needed, between input and output supplies. Figure 7-14 shows how this capacitance-switching power converter generates a negative voltage from a positive one. Figure 7-15 shows how to get the needed ±15 V from +5 V. Note the diode which is back biased and blocks flow back into the +5 V source since +15 V is on its cathode.

The dropout voltage for the ICL7663 and 7664 is just 0.1 V at 1 mA. Both regulators are shown in block diagram form in Fig. 7-16. Figure 7-17 shows the positive voltage regulator (7663) used in a conventional manner much as any voltage regulator IC would be used. Both this positive regulator and the negative regulator shown in Fig. 7-18 draw only 20 microamps while idling. These regulators are capable of 20 mA outputs and even at these elevated currents, their dropout voltages remain below that of the 723 or 109 voltage regulators. The generation of split supplies from a single supply is shown in Figs. 7-19 and 7-20. The capacitor, C_{ref}, is for timing purposes and if the capacitor is increased, the clock rate of the

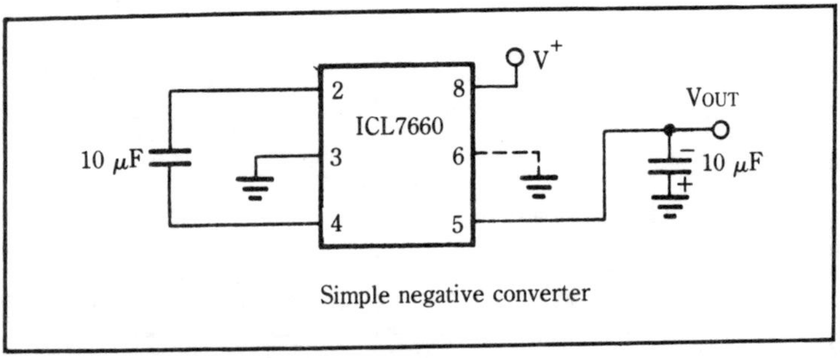

Fig. 7-14. The ICL7660 capacitance-switching principle. (From Intersil Data Book, July 1979; reprinted with permission.)

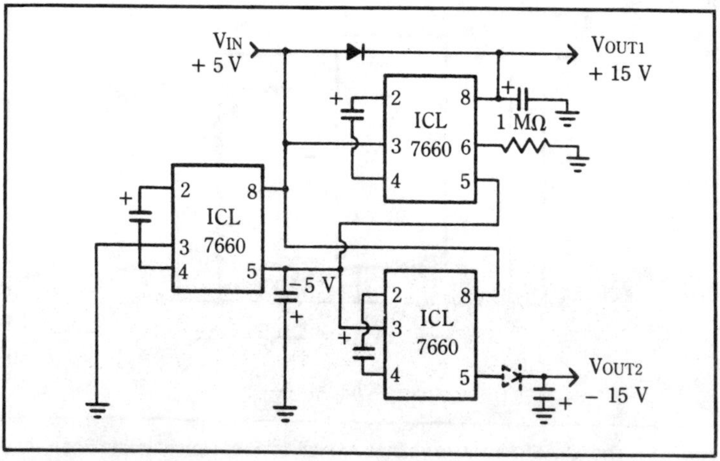

Fig. 7-15. Deriving ± 15 V from only + 5 V. (From Intersil Data Book, July 1979; reprinted with permission.)

ICL7660 will decrease. This also decreases power consumption. Remember in the CMOS discussion how power consumption in CMOS increases with clock frequency. Figure 7-21 illustrates the relative efficiency of the ICL7660 converter over the ordinary resistive voltage divider method.

If power efficiency is still a prime concern and the 20 mA capability is too small, you can use the circuit in Fig. 7-22A and derive −5 V to −6 V at 150 mA from an input of +10 V. This is

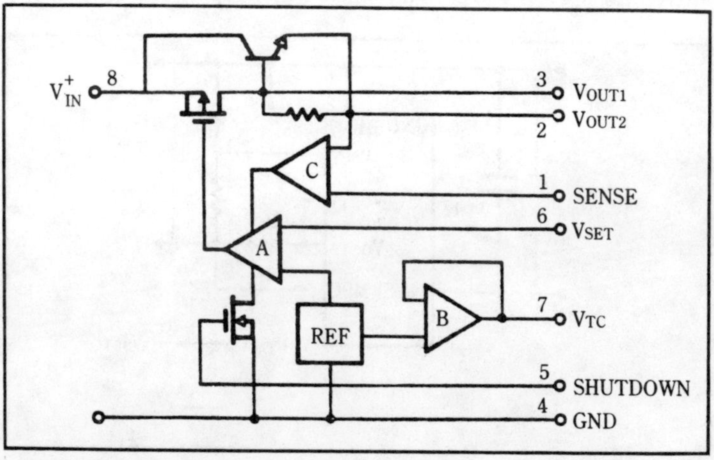

Fig. 7-16. The ICL7663/4 block diagram. (From Intersil Data Book, July 1979; reprinted with permission.)

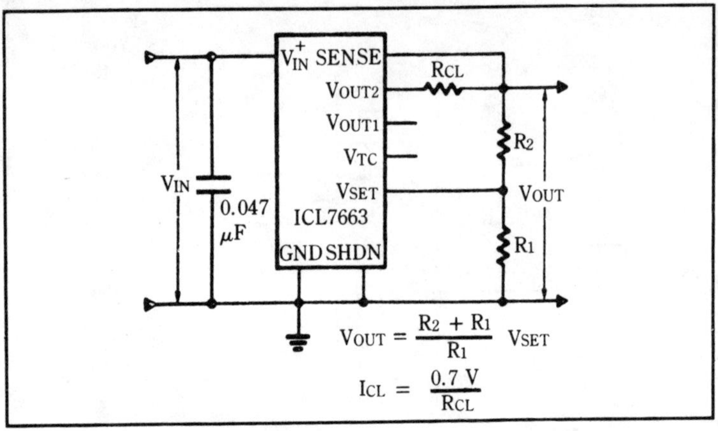

Fig. 7-17. The ICL7663 positive voltage regulator. (From Intersil Data Book, July 1979; reprinted with permission.)

accomplished by diverting charging currents from internal ICL7660 transistors and capacitors to external current-boosting components. Refer to Fig. 7-22B and note how capacitor C1 charges to the input voltage when transistors Q1 and Q3 are turned on and Q2 and Q4 are open. When Q2 and Q4 are turned on, part of the charge on Q1 is transferred to C2, developing a negative potential across C2. Subsequent charging and discharging cycles of C1 and C2 brings the voltage across C2 up to the input voltage level.

Under load and during the charge cycle of C1, C2 discharges partly into R_L. This loss is minimized by keeping the C1 charge/

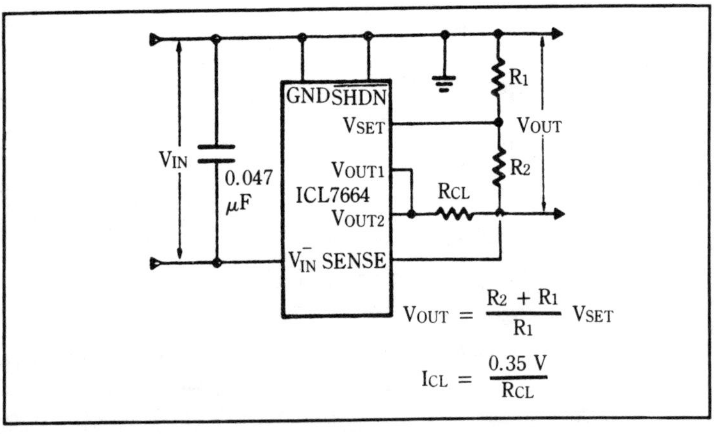

Fig. 7-18. The ICL7664 negative voltage regulator. (From Intersil Data Book, July 1979; reprinted with permission.)

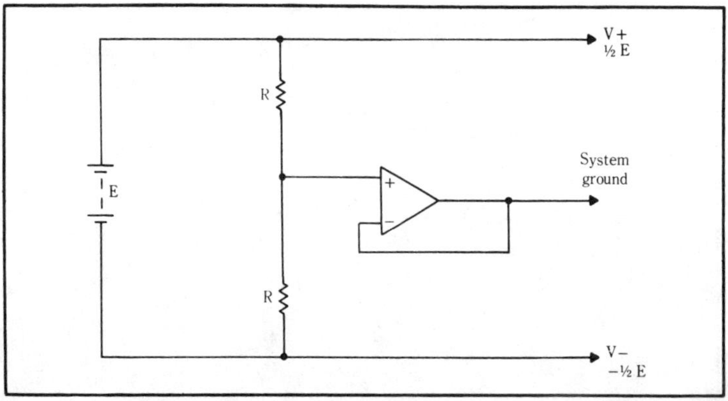

Fig. 7-19. Resistive voltage divider yielding split supplies from single supply. (From *Electronics*, July 3, 1980; reprinted with permission.)

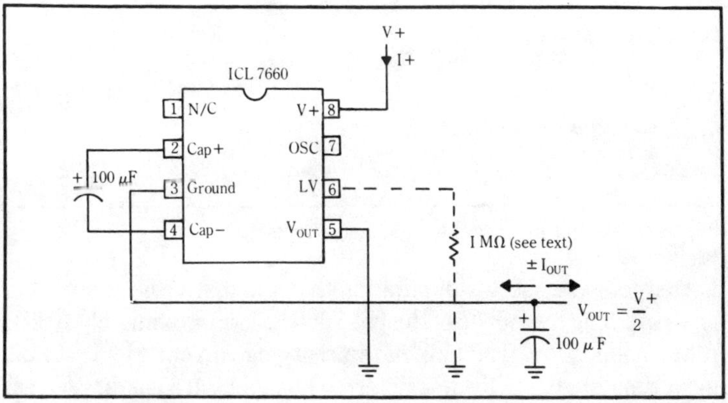

Fig. 7-20. Actual circuit hook-up using ICL7660 converter. (From *Electronics*, July 3, 1980; reprinted with permission.)

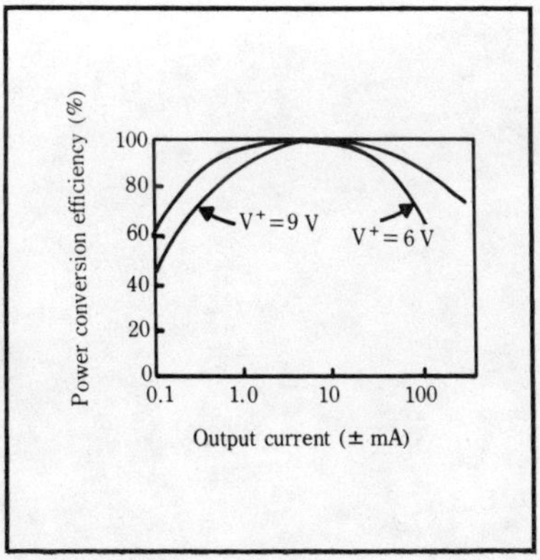

Fig. 7-21. Comparison of efficiency of simple resistive voltage divider vs the ICL7660 converter. (From *Electronics*, July 3, 1980, reprinted with permission.)

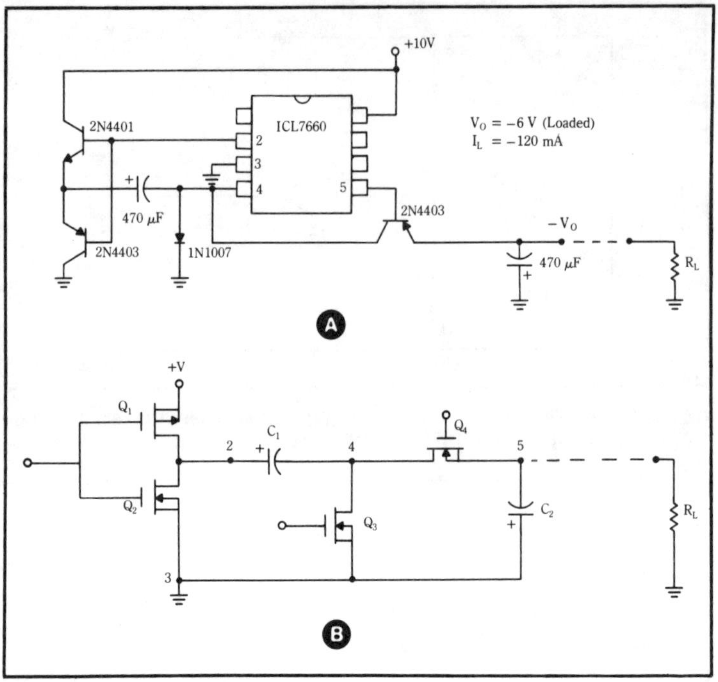

Fig. 7-22. A current-boosted ICL7660 application.

discharge cycle shorter in duration than the discharge time of C2. The switching frequency of the ICL7660 IC is approximately 10 kHz in this application. The bulk of the charging current to C1 and C2, and ultimately to the load, is diverted by bipolar transistors such as the three in the circuit. The diode is to absorb current surges, especially at turn on.

Chapter 8

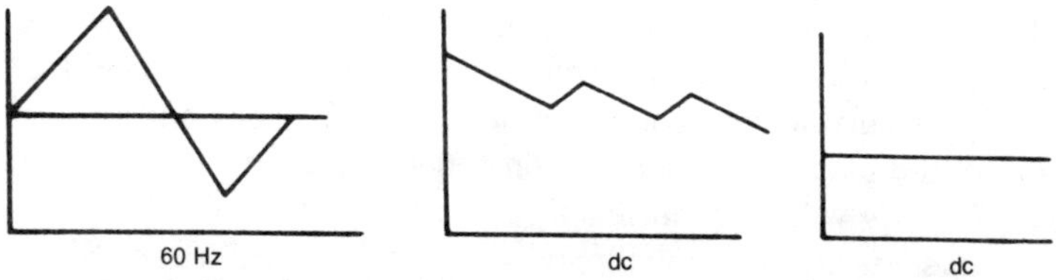

Thermal Considerations and Reliability Enhancement

This chapter is disproportionately larger than most of the others, but this is with good justification. In major electronic systems, reliability is paramount, and the prime cause of failures, the "Achilles heel," is the connectors. Connector pins break, get bent, or are improperly mated thereby making poor or no connection at all. This is not a theory or mere conjecture on my part. Years of collected failure data to determine military electronic system mean time between failures (MTBF) and military contracts more recently awarded on a life cycle cost basis (the more maintenance free an electrical system is, the more money the company will receive in the years subsequent to its deployment in the field) naturally has mandated very carefully kept, well-documented records substantiating failures and their causes. Connectors, though, are most solely under the control of a manufacturer of connectors, and we can do very little about them. Additionally, the smaller circuits you will probably be building from this book will fortunately not have that many interconnections.

However, the second most vulnerable cause of electronic system failures is a special concern to us. This is the power supply failure which is directly traceable to excessive internal component or ambient heat in approximately 85 percent of all cases.

The considerations in design, modification, and use of both voltage regulator ICs and electronic assemblies as well as the enclosure's design are, therefore, dealt with in considerable detail in this chapter. It is my resolute desire that you enjoy this book. To do so, the circuits presented in the first seven chapters must not only be made to work but they must also be made capable of functioning in a manner yielding years of reliable service. This chapter is crucial, and here is where the reliability will either be designed in or where the design will become a problem.

Let's resign ourselves to keep semiconductors, voltage regulator ICs (the subject of this book), their associated support elements (such as series-pass transistors), and the enclosure itself *cool.* The surrounding enclosure not only influences but can also impact the performance of the components housed within it. This very shortly will become readily apparent.

It is crucial for transistors, ICs, or thyristors (triacs and SRCs) to remain within their safe operating junction temperature ranges. Semiconductor junctions can withstand anywhere between 85° C and 200° C (germanium withstands from 85° C to 100° C maximum temperature, while silicon withstands from 150° C to 200° C maximum temperature). However, semiconductors operated at these elevated temperatures often experience diminished critical operating parameters. As an example, low-input bias op amps double the input bias current required for every 10° C rise in junction temperature. Also, note the typical power derating curve in Fig. 8-1 for a power transistor.

Both junctions in a transistor give off heat; however, since the collector-base junction is reverse biased, it naturally has higher resistance and produces more heat. In fact, the collector-base junction produces so much more heat than the transistor's forward-biased emitter-base junction that the heat generated by this junction is usually ignored. As a result the power dissipation equation is simplified and can be defined as follows.

$$P_D = I_C \times V_{CE}$$ **Equation 8-1**

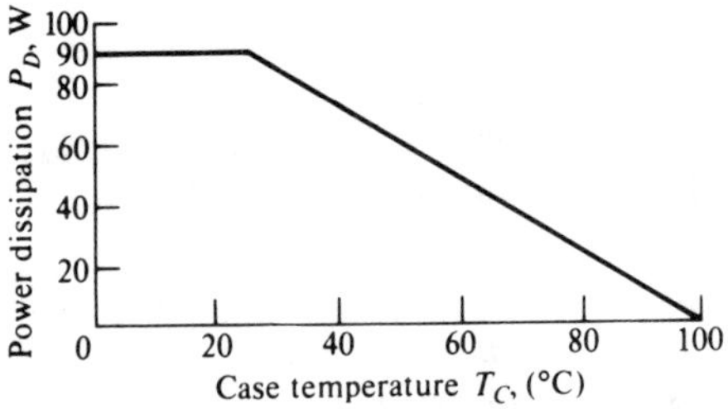

Fig. 8-1. Power dissipation versus case temperature or power derating curve for a transistor housed in a TO-3 case.

where P_D = power dissipation
I_C = collector current
V_{CE} = voltage from collector to emitter

ELECTRICAL-THERMAL ANALOGIES

Analogies are often helpful in visualizing an abstract phenomenon. As an example, high school electronics classes compare a water tower to electricity. In this comparison, voltage is defined as the "push" or potential (height) of the water within this tower. Current is the flow of water and resistance in the tower's faucet, which impedes or constricts flow to varying degrees, depending on the faucet setting.

An analogy can also be drawn between heat transfer and electrical parameters. The flow (transfer of heat) is "current" the "voltage" is the heat differential between the hottest object (the transistor's junction) and the coolest object (the free-standing ambient air). Each thermal resistance from the transistor's junction to its case, to the heatsink, and eventually to the free-standing ambient air is analogous to electrical resistances in series; therefore, thermal resistances are added together to determine the total thermal resistance.

Figure 8-2 illustrates the similarity of heat to electricity. The dual subscript on a thermal resistance, such as θ_{SA}, means resistance of the sink (S)-to-ambient air (A). Table 8-1 illustrates the unit symbols in both the electrical and thermal models.

The basic thermal relationship is:

$$T = P_D \theta \qquad \textbf{Equation 8-2}$$

where T = temperature rise
P_D = power dissipation
θ = thermal resistance

It becomes obvious from this equation that thermal resistance θ must be minimized.

HEATSINKS

Power transistors and voltage regulators with 1 A or greater ratings usually come in larger, more massive metal cases so that they can radiate more heat. The case size is often inadequate for the power to be dissipated so a heatsink or a metal fitting is placed over

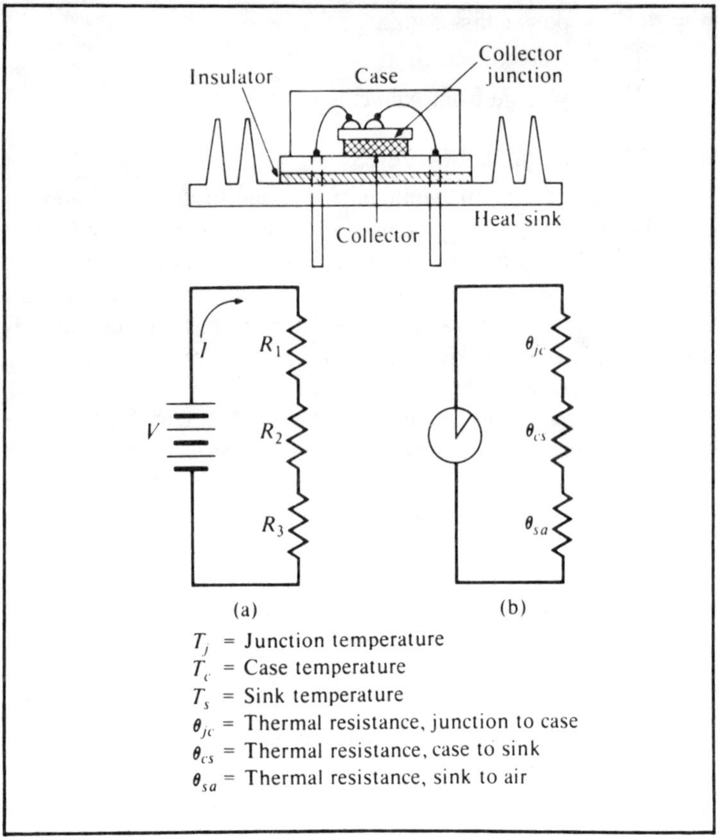

Fig. 8-2. An electrical-thermal analogy.

this case to bolster the thermal mass, lower the thermal resistance, and provide added surface area allowing additional heat to more effectively radiate from this surface.

Naturally, the basic purpose of a heatsink is to transfer heat away from the semiconductor, and this heat is transferred or flows

Table 8-1. Electrical-Thermal Units.

Electrical			Heat		
Quantity	**Unit**	**Symbol**	**Quantity**	**Unit**	**Symbol**
Voltage	V	V	Temperature rise	°C	T
Current	A	I	Heat flow	W	P_D
Resistance	Ω	R	Thermal resistance	°C/W	θ

Table 8-2. Emissivities of Various Common Surfaces. (Courtesy National Semiconductor.)

Surface	Emissivity, E
Polished Aluminum	0.05
Polished Copper	0.07
Rolled Sheet Steel	0.66
Oxidized Copper	0.70
Black Anodized Aluminum	0.7—0.9
Black Air Drying Enamel	0.85—0.91
Dark Varnish	0.89—0.93
Black Oil Paint	0.92—0.96

from one medium to another by means of one or more of the three heat transfer methods, which are as follows:

- Conduction is the most effective method for moving heat from the semiconductor junction to the case and from the case to the heatsink. This is where the heat conducts through the entire conductor.
- Convection is the most effective method for moving heat from case or heatsink to ambient air. This is surface conduction from a surface to a fluid or to air.
- Radiation is the most effective method for transferring heat from radiating heatsink fins. It is the principle based on the temperature difference between radiating and adjacent objects or space.

As just stated, heat transfer by radiation occurs from a surface. *Emissivity* is the term describing this effect, and Table 8-2 illustrates some typical emissivities of various common surfaces. The greater the emissivity, the better the ability to radiate heat. Note how black oil-painted heatsinks radiate most effectively. Commercially available heatsinks are almost exclusively painted black.

Heatsink Material

The two most common materials for heatsinks are copper and aluminum. Copper has a thermal conductivity four times that of aluminum, but it is much more expensive. Aluminum, therefore, is the most predominant heatsink material, although, if weight is a concern, magnesium can be used, or if insulation is a problem, beryllium oxide (BeO) can be used. It is an excellent electrical insulator.

Very often, power transistors such as the ones packaged in TO-3 or TO-66 cases ("TO" stands for transistor outline and defines the physical case shape and size) have their collectors soldered directly to the cases. This presents a shorting problem if the case is soldered to the chassis to aid in heat dissipation.

Insulating Washers

An insulating washer (refer to Fig. 8-3) with minimal thermal resistance is needed. There are three types of electrical insulating washers that are good heat conductors. These are mica, anodized aluminum, and beryllium oxide. Mica is the most common. Anodized aluminum is fairly widely used, but its surface must not be scratched or an electrical short will result. Beryllium oxide performs better than mica or anodized aluminum, but the powdered form is toxic and so are its fumes.

Custom-Made Heatsinks

For the do-it-yourself enthusiast in electronics who wants to make his own heatsinks, Fig. 8-4 and Fig. 8-5, nomograms of metal types and shapes and fin effectiveness, will help along with the following discussion. You may have access to some surplus ex-

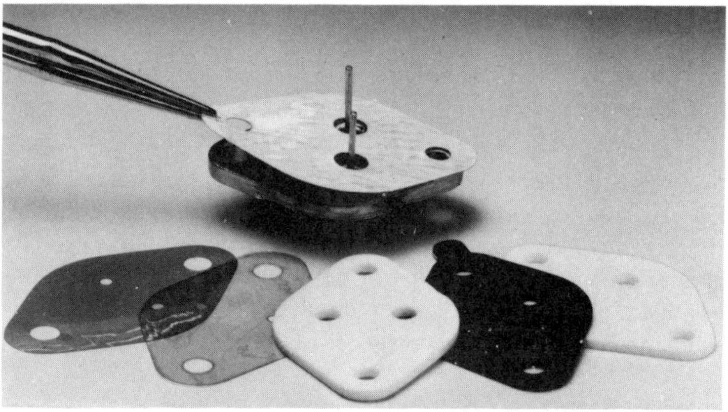

Insulator	Thermal resistance, °C/W	
	Dry	Silicon grease
None	0.2	0.1
Teflon	1.45	0.8
Mica	0.8	0.4
Anodized aluminum	0.4	0.35

Fig. 8-3. An insulating washer.

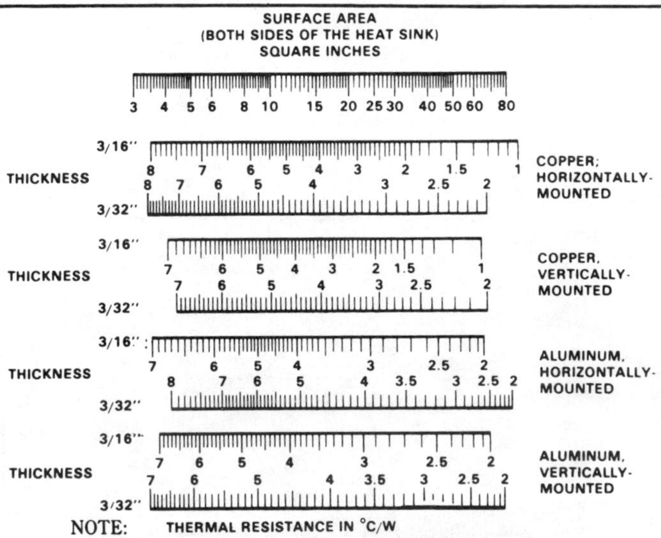

NOTE:

To determine either area required or thermal resistance of a given area, draw a vertical line between the top (or area) scale down to the material of interest. To determine the required surface area for a given material with a known thermal resistance, draw a vertical line from the appropriate scale up to the surface area scale.

Fig. 8-4. A nomogram of thermal effectiveness vs metal types and thicknesses.

truded copper or aluminum or just may want to attach a single flat square or rectangular piece of metal to a voltage regulator or power transistor and have wondered if it will do the job. A good question that deserves explanation!

Before making any calculations, be certain of the semiconductor's case. Specifically, if the semiconductor's case is a TO-220 or TO-202 metal tab, look at its data sheet to see if the metal tab is the collector on a transistor or the output pin on a voltage regulator, or if it is the emitter which might be grounded or the ground pin on a voltage regulator. If the metal surface of the semiconductor's case has voltage on it, paint the heatsink or chunk of metal you are about to fashion into a heatsink, so that it will be electrically insulated. A mica insulating wafer or washer may also be used but this adds a small thermal resistance. Now, let's see if the heatsink will do the job.

Heatsink Calculations

Let's go through a typical heatsink calculation. Equation 8-2, a

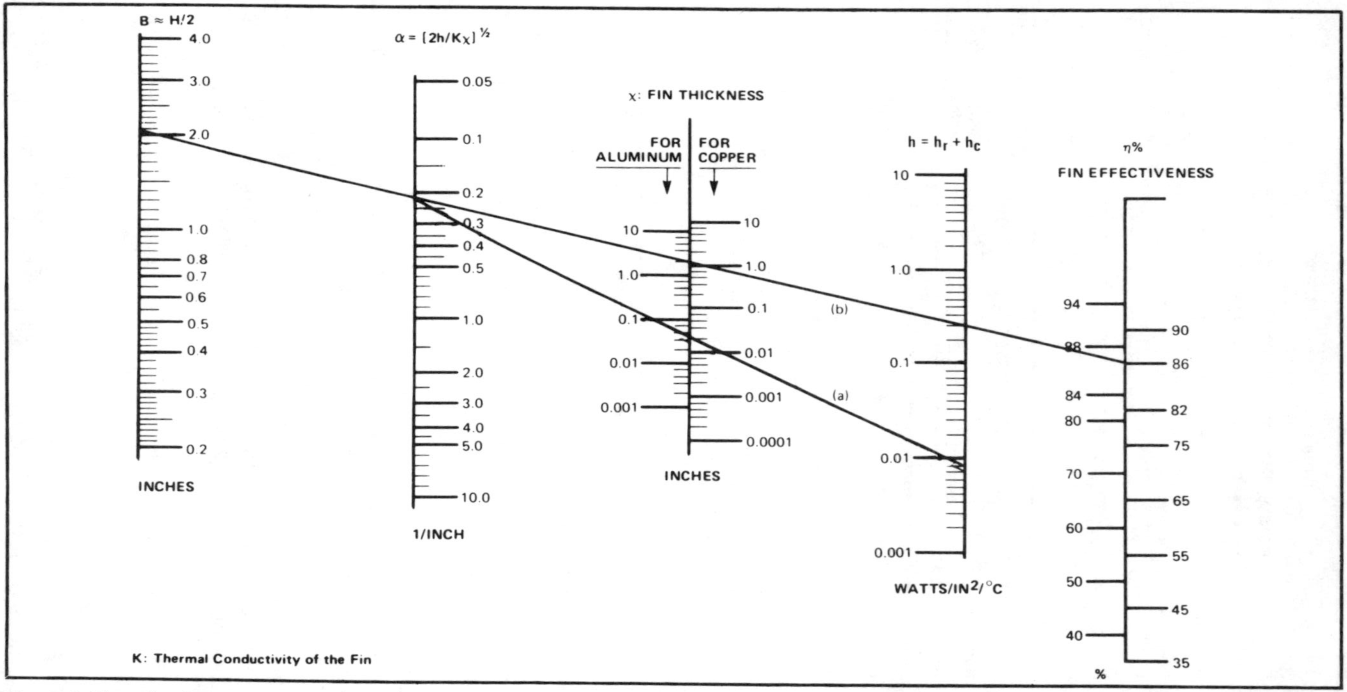

Fig. 8-5. Fin effectiveness nomogram for symmetrical, flat, uniformly thick, vertically mounted fins. (Courtesy National Semiconductor.)

generalized equation of temperature differential (rise), power dissipation, and thermal resistance, can be rewritten as :

$$P_{D(MAX)} = \frac{T_{J(MAX)} - T_{A(MAX)}}{\theta_{JA}}$$ **Equation 8-3**

where $T_{J(MAX)}$ = maximum junction temperature in ° C
$T_{A(MAX)}$ = maximum ambient air temperature in ° C
θ_{JA} = junction-to-air thermal resistance

This equation states that the maximum power dissipated is equal to the maximum junction temperature of the transistor or regulator minus the maximum ambient or outside free-standing air temperature. The difference in the two temperatures is divided by the thermal resistances from the junction (J) to the ambient air (A). These thermal resistances are analogous to electrical resistances in series, so they are added together. Stated mathematically, $\theta_{JA} = \theta_{JC} + \theta_{CS} + \theta_{SA}$; the thermal resistance from junction to air (θ_{JA}) equals the sum of the thermal resistances junction-to-case (JC), case-to-heatsink (CS) and heatsink-to-air (SA).

Quite simply, if the $P_{D(MAX)}$ of Equation 8-3 is exceeded, then a heatsink is required! Typically, a TO-3 case will withstand 2.8 W, and a TO-220 power tab with a flat rectangular body and three parallel leads coming out the end separated by 0.1 inch each will withstand 1.8 W. The smaller but identically shaped TO-202 package will handle 1.56 W, while the very small TO-39 and TO-92 packages handle only about two-thirds of a watt.

Let us assume the following in our heatsink calculations:

Maximum ambient temperature 60° C
Maximum junction temperature 125° C
Maximum output current 800 mA
Maximum input-output differential 10 V

If we use a 7800 series voltage regulator with a TO-220 case, the θ_{JC} = 5° C/W; this is obtained from the data sheet. The θ_{CS} and θ_{SA} must be solved for by combining Equation 1 and Equation 3.

$$\theta_{JA} = \theta_{JC} + \theta_{CS} + \theta_{SA} = \frac{T_J - T_A}{P_D}$$ **Equation 8-4**

$$= \frac{125° \text{ C} - 60° \text{ C}}{.8 \text{ A} \times 10 \text{ V}} = \frac{65° \text{ C}}{8 \text{ watts}} = 8.13° \text{ C/W}$$

where θ_{JC} = junction-to-case thermal resistance
θ_{CS} = case-to-heatsink thermal resistance
θ_{SA} = heatsink-to-air thermal resistance
T_J = junction temperature
T_A = ambient air temperature

Let us solve the far right side of the equation first, $T_J - T_A = 65°$ C. The denominator is merely the 800 mA output current times 10 V, the input-output voltage differential. The three thermal resistances, then, sum to 8.13° C/W. From the data sheet we already know that the junction-to-case thermal resistance is 5° C/W so the other two remaining resistances must add up to 3.13° C/W. With the use of silicon thermal grease, we can approximate with some assurance from the Table in Fig. 8-3 that the case-to-sink resistance is 0.13° C/W, so this leaves the $\theta_{SA} = 3°$ C/W. Refer to Fig. 8-4 and note that a 22 square-inch piece of aluminum 3/16 inch thick with vertically bent ends will meet this requirement. The note at the bottom of Fig. 8-5 explains how to use the nomograph.

Calculating the effectiveness of a heatsink with fins is much more involved than with a heatsink of square or rectangular metal as was just done. The heatsink-to-ambient air thermal resistance is expressed as:

$$\theta_{SA} = \frac{1}{2H^2 \eta (h_c + h_r)} \text{ ° C/W}$$ **Equation 8-5**

where H = vertical height of fins in inches
η = fin effectiveness factor
h_c = convection heat transfer coefficient
h_r = radiation heat transfer coefficient

$$h_c = 2.21 \times 10^{-3} \left(\frac{T_S - T_A}{H} \right)^{1/4} \text{ W/in}^2 \text{ ° C}$$ **Equation 8-6**

$$h_r = 1.47 \times 10^{-10} \text{ E} \left(\frac{T_S + T_A}{2} + 273 \right)^3 \text{ W/in}^2 \text{ ° C}$$ **Equation 8-7**

where T_S = temperature (° C) of heatsink at regulator mounting
T_A = temperature (° C) of ambient air
E = surface emissivity (refer to Table 8-2)
H = vertical height of fins in inches

Now with the aid of Fig. 8-5 and the formulas, let's try to design a finned heatsink for a 5 V regulator. Let's assume we are using a black anodized 1/16-inch thick piece of aluminum and we will have symmetrical square heatsink fins. Our desired heatsink will have a thermal resistance of 4° C/W. An LM340T-5 is used under the following realistic conditions:

T_J = 125°C V_{OUT} = 5 V
T_A = 60° C I_{OUT} = 800 mA
V_{IN} = 15 V

We determine T_S by adapting Equation 8-3; substitute in Equation 8-1 for P_D, and rearrange to get $IV\theta_{JS} = T_J - T_S$. V equals $V_{IN} - V_{OUT}$. We then get (15 V-5 V) 800 mA × 4° C/W = 125° C – T_S, which solves to 32° C = 125° C – T_S. T_S therefore equals 125° C – 32° C, or 93° C. Next h_c is solved for using a 4¼-inch high fin.

$$h_c = 2.21 \times 10^{-3} \left(\frac{93° \text{ C} - 60° \text{ C}}{4.25 \text{ in}}\right)^{1/4} \qquad \textbf{Equation 8-8}$$

$$= 3.7 \times 10^{-3}$$

And h_r is solved for with E = .9, which is found in Table 8-2.

$$h_r = 1.47 \times 10^{-10} \times 0.9 \left(\frac{93° \text{ C} + 60° \text{ C}}{2 \text{ in}}\right)^3 \qquad \textbf{Equation 8-9}$$

$$= 5.6 \times 10^{-3}$$

Next, add h_c to h_r to get $h = 9.3 \times 10^{-3}$. Now we find η from Fig. 8-5. Draw the first line on the nomograph in Fig. 8-5 from h through the fin thickness (remember, we are using .0625″ (1/16″) aluminum) to find α. Next, determine B, which is approximately the height of the fins divided by 2, or 2.1 inches. Draw the second line from this value of B through $\alpha = .38$ and extend it all the way over to η which is a determination of fin effectiveness. Now that we have found η graphically to be 86 percent, the equation for θ_{SA} can be solved.

If the solution of θ_{SA} is too large, this indicates an inadequate amount of heat-radiating ability and the fin height should be increased. If the equation yields a number that is less than the designed thermal resistance, 4° C/W in this case, the heatsink is too large. However, too large a heatsink is not a problem because it only provides an additional thermal safety factor. If physical case restraints prevent use of a large heatsink ,or if economics prevent its use, you can go to either a copper heatsink or a smaller one or reduce the V_{IN} minus V_{OUT} differential. All of these will help.

Returning to the derivation of θ_{SA}, its h_r constituent term can also be improved upon from the 0.9 we used in the form of black anodized aluminum. Table 8-2 indicates that black oil paint can have an E of 0.96. The preceding derivation of θ_{SA} assumed that the heatsink was vertically mounted and that the fins were symmetrically square. If horizontal mounting is used, derate h_c by multiplying it by 0.8. Horizontal mounting of any heatsink is not a practice whereby its maximum effectiveness can be realized. If a finned-heatsink is mounted horizontally, this is even worse. But assuming that the confines of the components in a case necessitates this, the "one side only" effective finned heatsink should have its η multiplied by 0.5 and its h_c derated by being multiplied by 0.94.

With the values used thus far, η has been graphically determined to be 0.86. Substituting our numbers in Equation 8-5 we obtain the following:

$$\theta_{SA} = \frac{10^3}{2 \times 18 \times 0.86 \times 9.3} = 3.47° \text{ C/W} \qquad \textbf{Equation 8-10}$$

Incidentally, the 10^{-3} was changed to 10^3 by moving it to the numerator and changing the sign of the exponent. The value of 3.47° C/W is not exactly the desired 4.00° C/W we initially sought, but gives a margin of safety. Remember, when doing it yourself, these formulas apply to round and square symmetrical fins. Rectangular 2:1 aspect ratio (length-to-height) fins may also be used with the deratings applied. Now you can build them yourself and know they will do the job! (Table 8-3 shows some typical θ_{SA} values per case size in commercially available heatsinks.) By following Fig. 8-4 and Fig. 8-5 you can build your own heatsink.

Heatsink User Guidelines

After addressing the subject of homemade heatsinks, let us now examine some heatsink user guidelines.

Table 8-3. Commercially Available Heatsinks. (Courtesy National Semiconductor.)

No attempt has been made to provide a complete list of all heatsink manufacturers. This list is only representative.

	Manufacturer & Type
For TO-202 Packages	
12.5 - 14.2	Staver V4-3-192
13	Staver V5-1
15.1 - 17.2	Staver V4-3-128
19	Thermalloy 6106 Series
20	Staver V6-2
25	Thermalloy 6107 Series
37	IERC PA1-7CB with PVC-1B Clip
40 - 42	Staver F7-3
40 - 43	Staver F7-2
42	IERC PA2-7CB with PVC-1B Clip
42 - 44	Staver F7-1
For TO-220 Packages	
4.2	IERC HP3 Series
5 - 6	IERC HP1 Series
6.4	Staver V3-7-225
6.5 - 7.5	IERC VP Series
8.1	Staver V3-5
8.8	Staver V3-7-96
9.5	Staver V3-3
10	Thermalloy 6032, 6034 Series
12.5 - 14.2	Staver V4-3-192
13	Staver V5-1
15	Thermalloy 6030 Series
15.1 - 17.2	Staver V4-3-128
16	Thermalloy 6106 Series
18	Thermalloy 6107 Series
19	IERC PB Series
20	Staver V6-2
25	IERC PA Series
26	Thermalloy 6025 Series
For TO-92 Packages	
30	Staver F2-7
46	Staver F5-7A, F5-8-1
50	IERC RUR Series
57	Staver F5-7D
65	IERC RU Series
72	Staver F1-7
85	Thermalloy 2224 Series

θ_{SA} Approx[1] (°C/W)	Manufacturer & Type
For TO-5 Packages	
12	Thermalloy 1101, 1103 Series
12 - 16	Wakefield 260-5 Series
15	Staver V3A-5
22	Thermalloy 1116, 1121, 1123 Series
22	Thermalloy 1130, 1131, 1132 Series
24	Staver F5-5C
26 - 30	IERC Thermal Links
27 - 83	Wakefield 200 Series
28	Staver F5-5B
30	Thermalloy 2227 Series
34	Thermalloy 2228 Series
35	IERC Clip Mount Thermal Link
39	Thermalloy 2215 Series
42	Staver F5-5A
45 - 65	Wakefield 296 Series
46	Staver F6-5, F6-5L
50	Thermalloy 2225 Series
50 - 55	IERC Fan Tops
51	Thermalloy 2205 Series
53	Thermalloy 2211 Series
55	Thermalloy 2210 Series
56	Thermalloy 1129 Series
58	Thermalloy 2230, 2235 Series
60	Thermalloy 2226 Series
68	Staver F1-5
72	Thermalloy 1115 Series

θ_{SA} Approx[1] (°C/W)	Manufacturer & Type
For TO-3 Packages	
0.4 (9" length)	Thermalloy (Extruded) 6590 Series
0.4 - 0.5 (6" length)	Thermalloy (Extruded) 6660, 6560 Series
0.56 - 3.0	Wakefield 400 Series
0.6 (7.5" length)	Thermalloy (Extruded) 6470 Series
0.7 - 1.2 (5 - 5.5" length)	Thermalloy (Extruded) 6423, 6443, 6441, 6450 Series
1.0 - 5.4 (3" length)	Thermalloy (Extruded) 6427, 6500, 6123, 6401, 6403, 6421, 6463, 6176, 6129, 6141, 6169, 6135, 6442 Series
1.9	IERC E2 Series (Extruded)
2.1	IERC E1, E3 Series (Extruded)
2.3 - 4.7	Wakefield 600 Series
4.2	IERC HP3 Series
4.5	Staver V3-5-2
5 - 6	IERC HP3 Series
5.2 - 6.2	Thermalloy 6103 Series
5.6	Staver V3-3-2
5.8 - 7.9	Thermalloy 6001 Series
5.9 - 10	Wakefield 680 Series
6	Wakefield 390 Series
6.4	Staver V3-7-224
6.5 - 7.5	IERC UP Series
8	Staver V1-5
8.1	Staver V3-5
8.8	Staver V3-7-96
8.8 - 14.4	Thermalloy 6013 Series
9.5	Staver V3-3
9.5 - 10.5	IERC LA Series
9.8 - 13.9	Wakefield 630 Series
10	Staver V1-3
13	Thermalloy 6117

Staver Co, Inc: 41-51 N. Saxon Ave, Bay Shore, NY 11706
IERC: 135 W. Magnolia Blvd, Burbank, CA 91502
Thermalloy: PO Box 34829, 2021 W. Valley View Ln, Dallas TX
Wakefield Engin Ind: Wakefield MA 01880

[1] All values are typical as given by mfgr. or as determined from characteristic curves supplied by mfgr.

- Avoid placing voltage regulators next to heat generating components like power resistors.
- When using smaller package regulators like the TO-5, TO-39, or TO-92 cases, keep the lead lengths to a minimum and use the largest possible PC (printed circuit) board copper connector runs to provide a heat dissipation path for the regulator.
- Make certain that the heatsink to case interface is very flat. With larger heatsinks, this becomes more difficult, so for good thermal conduction, a thin layer of silicon thermal grease is used. Three leading types of thermal grease are Dow Corning 340, General Electric 662, and Thermalloy's Thermacote. Metallic-oxide-filled silicon grease placed between two surfaces fills in any air gaps or scratches on the two metal surfaces. A typical transistor mounted dry with a 0.2° C/W resistance can have this resistance halved with use of this joint compound. Recall that we used a 0.13° C/W rating in our former calculation. Figure 8-6 shows some joint compound and a variety of heatsinks.
- Negative voltage regulators in particular require that the heatsink and metal voltage regulator case be electrically insulated from one another. This is accomplished with either a fiberglass or mica 0.003- to 0.005-inch thick insulating washer. This naturally increases thermal resistance but this can be partially offset by applying thermal grease to each side of the washer.
- When mounting a regulator on a heatsink with fins, it is best to orient the fins in the vertical plane.
- If it is ever necessary to bend the regulator's leads, be very careful, because this is very detrimental to reliability.

OTHER HEAT VENTING MEANS

There are instances in which heatsinks are still inadequate. They can pull heat away quickly, but the ambient or surrounding air temperature can rise or edge up in time if the power supply is enclosed in a poorly ventilated cabinet. This is where a fan or blower comes to the rescue. However, the two are not one and the same.

Fans vs. Blowers

Blowers usually consist of a wheel that revolves to displace air and they are most efficient while operating near their maximum static (non-moving) pressure. Fans (refer to Fig. 8-7) usually consist of propeller blades and operate best when moving large volumes of low-pressure air.

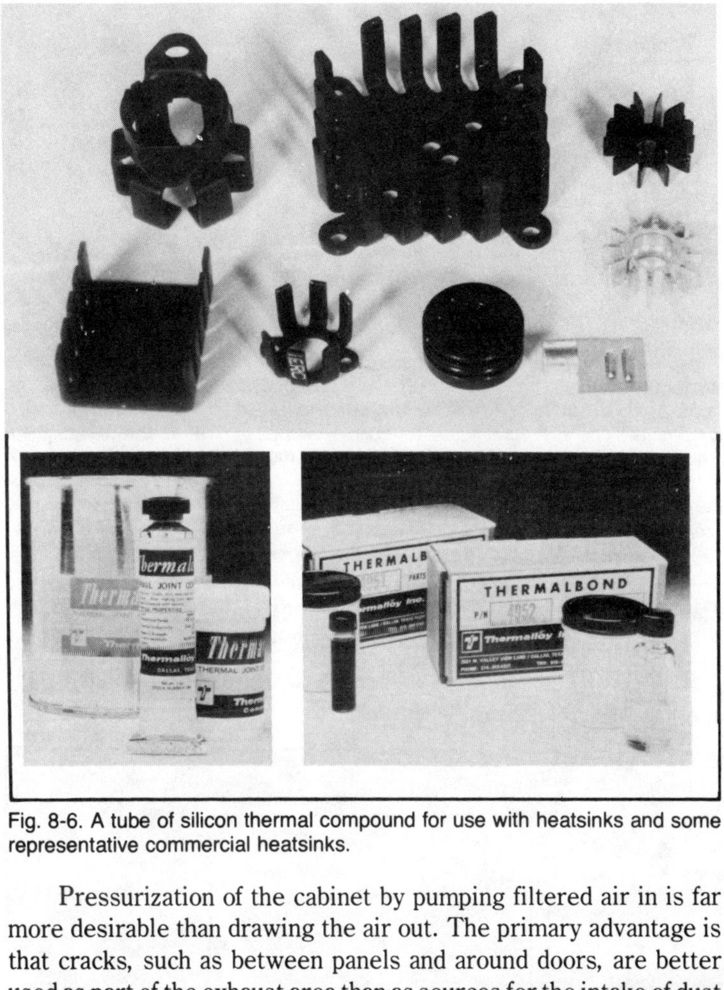

Fig. 8-6. A tube of silicon thermal compound for use with heatsinks and some representative commercial heatsinks.

Pressurization of the cabinet by pumping filtered air in is far more desirable than drawing the air out. The primary advantage is that cracks, such as between panels and around doors, are better used as part of the exhaust area than as sources for the intake of dust and dirt. The presence of air-borne particulate material compounds the problem of controlling the temperature of electronic equipment.

There are four methods of cooling electronic equipment housed in a cabinet. These are:

- Forced-convection filtered air (Fig. 8-8)
- Air-to-air heat exchangers
- Air-to-water heat exchangers and
- Specially packaged air conditioning units.

The last three are for large pieces of equipment and, therefore, are not covered here.

Fig. 8-7. A typical fan used for ventilation purposes. Installation can be made for either pressurization or evacuation. Units featuring ball-bearing motors can be mounted in either vertical or horizontal positions, inside or outside the electronic enclosure. These mechanically reversible fans have a deep throat orifice for improved airflow and cooling efficiency. Four models provide air deliveries of 275, 390, 725 and 1200, CFM. Accessories available for each model include a heavy gauge stainless steel grille, a permanent washable type filter, and a zinc luster-plated propeller guard designed to fit either side of fan and meet normal OSHA requirements. (*Courtesy McLean Engineering*).

Design Guidelines for Forced-Air Cooling

There are five design guidelines to follow in the proper forced-convection filtered air-cooling of the interior of an equipment cabinet. They are as follows:

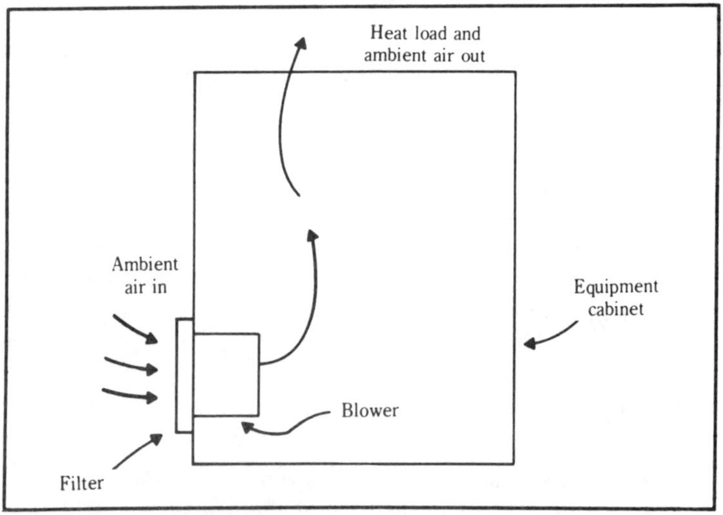

Fig. 8-8. An illustration depicting forced-convection filtered air cooling. This technique uses ambient air to cool electronic control equipment. Incoming air is filtered before being blown throughout the equipment cabinet.

1. The cross-section area of the air stream throughout the flow path should be at least equal to the effective area of the air intake.

2. The exhaust area must be located downstream from the heat-producing elements.

3. Use of a baffle works best when an exceedingly hot component has this baffle to channel a small volume of air across this location at a high velocity.

4. Ducts are used to maintain a more even cooling effect throughout the cabinet. If the maintenance of an even temperature from the top to bottom of the cabinet is important, place ducts along the sides of the cabinet.

The importance of even cooling may be dramatically underscored by the following excerpt from the *1978 Raytheon Linear Handbook* describing their precision analog multiplier. "The scale factor is sensitive to temperature gradients across the chip in the lateral direction. Where possible, the package should be oriented such that sources generating temperature gradients are located physically on the line of thermal symmetry. This will minimize scale-factor error due to thermal gradients."

5. Blowers and especially fans can cause vibration so look for neoprene vibration isolators or similar devices built into the fan if this is a concern.

Fan Air-Flow Nomograph

If you decide you need a fan, the following two formulas will enable you to determine what capacity fan in volume of air flow at a specified rate is required.

$$\text{Volume of air at the Inlet} = \frac{3.17\ (\text{power})\ 1.25}{T\ (°\ F)} \qquad \textbf{Equation 8-11}$$

$$\text{Volume of air at the Inlet} = \frac{1.76\ (\text{power})\ 1.25}{T\ (°\ C)} \qquad \textbf{Equation 8-12}$$

The power to be dissipated is expressed in watts and the temperature is the average temperature rise within the cabinet, expressed in Equation 8-11 as degrees Fahrenheit and Equation 8-12 as degrees Celsius. These two formulas as well as the nomograph in Fig. 8-9 have a built in safety factor of 25 percent (note the 1.25 for 125 percent). If only a quick approximation is required, the nomograph may be used.

Let's go through a typical design example assuming we have an average cabinet. Let's use the formulas as well as the nomograph and strike a line through it and see how close the nomograph comes to the exact calculated value.

Assume we have exactly 200 watts of power to be dissipated and we can only stand a 10° F rise in temperature within the cabinet. Equation 8-11 yields [(3.17) (200 watts) (1.25)]/10 = 79.25 cubic feet per minute. Note in Fig. 8-9 that the nomograph indicated

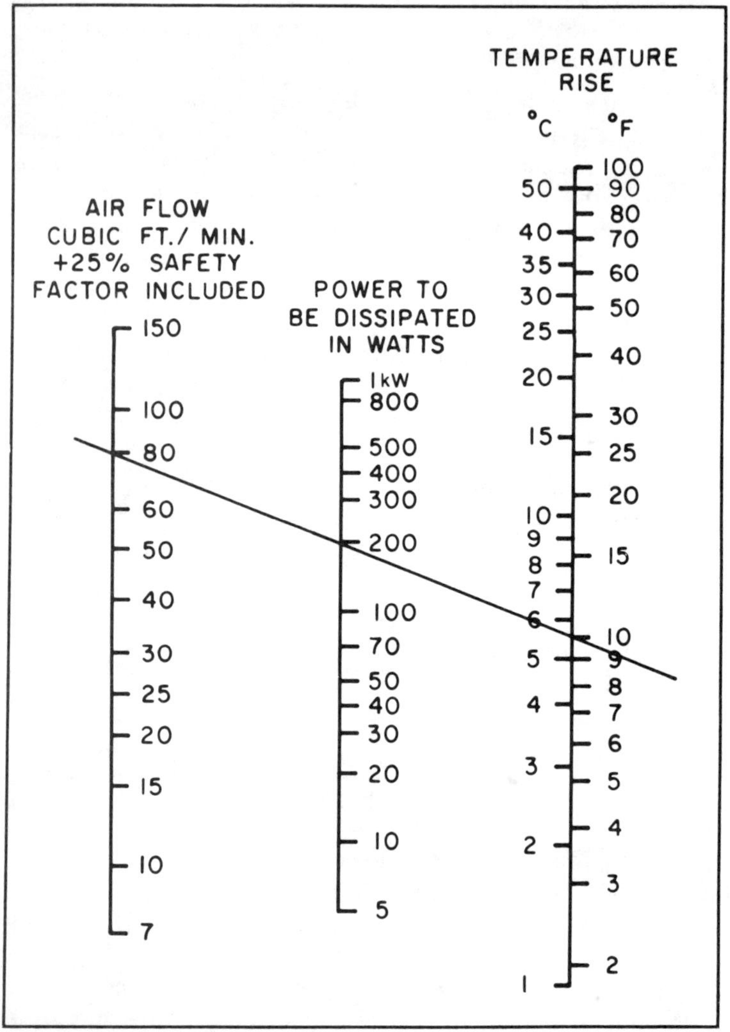

Fig. 8-9. Fan Air-Flow Nomogram.

approximately 80 cubic feet per minute which is very close to 79.25, the value derived by actual calculation.

Thermoelectric Devices

In addition to forced air venting, thermoelectric devices such as Peltier and Thomson-Joule devices may be used to pull heat away from a specific small area in the cabinet or a critical semiconductor which must remain cool.

There are four basic physical phenomena associated with thermoelectric devices. These four phenomena are:

- The Seebeck effect is the EMF that exists when two dissimilar conductors are connected and have their junctions maintained at different temperatures. This is the basis of thermocouples.
- The Thomson effect is the heating or cooling effect that takes place in a homogeneous conductor when an electric current passes in the direction of the temperature gradient.
- The Joule effect is when an electric current passes through a conductor (which is isothermal or maintains the same temperature throughout) and generates heat. This is called Joule heat.
- Last, the Peltier effect, which was named in honor of Jean C. A. Peltier, who in 1934, discovered that the passing of electric current through the junction of two dissimilar materials causes a

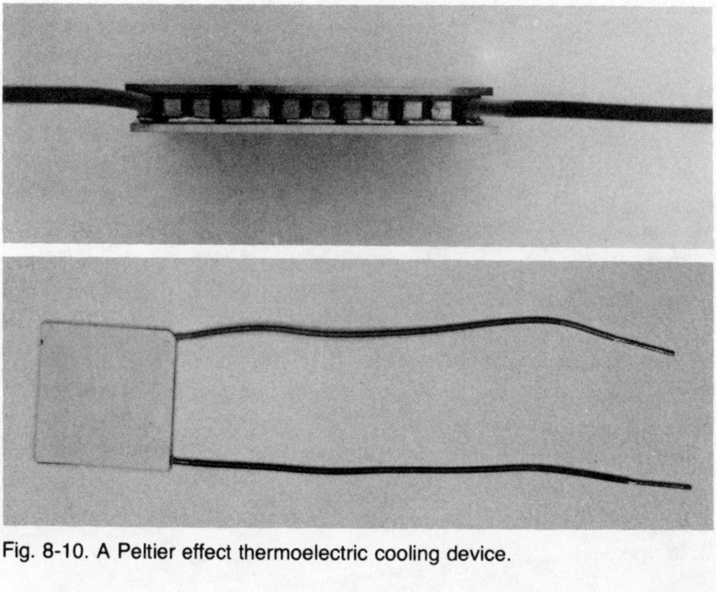

Fig. 8-10. A Peltier effect thermoelectric cooling device.

cooling effect when passed in one direction and a heating effect when the direction of current flow is reversed.

The Seebeck, Peltier, and Thomson effects are all reversible phenomena. The Peltier effect is the one most useful in power supply design because of its ability to literally draw heat away from a semiconductor, although the Thomson effect also has some limited applications in these areas and will be discussed shortly.

Companies today, such as Cambion produce Peltier cooling devices (refer to Fig. 8-10) which may be attached directly to surfaces of heat producing semiconductors. The effect of heat being drawn away in proportion to the current passing through the cooling device is what makes this device so effective. These devices though are somewhat expensive, with a typical price of $15 for a device large enough to handle a TO-3 power transistor.

Another interesting thermoelectric device is the Joule-Thomson cooler, as shown in Fig. 8-11. This "micro-miniature refrigerator", as its manufacturer refers to it, cools a wide range of IR (infrared) and millimeter wave detectors down to 77° K, or Kelvin, which is the temperature of liquid nitrogen (N_2). Zero degrees Kelvin is *absolute zero* (where all molecular motion stops) this is equal to −273.15° C or −459.76° F. (One degree Kelvin equals in magnitude one degree Celsius so to convert Kelvin to Celsius, just subtract 273.15 degrees from it. So 77° K = 77−273.15 = −196.15° C or −321.07° F.) These 1.5 cm (0.6 inch) in diameter devices can cool some detectors to 80° K (−193.15° C) in just one second.

The Piezoelectric Fan

Another interesting device used to accomplish cooling is the piezoelectrically driven fan designed for PC board use. This device, manufactured by Piezo Electric Products (refer to Fig. 8-12) is available in either 50 or 60 Hz and even 240 V as well as 120 V models. The device is used for spot cooling and uses about 1/15 the power of a fan accomplishing the same cooling effect. Its reliability is very high because of flexible metal strap blades laminated to thin-sheet piezoceramic elements. The large mechanical vibrations that result when an ac voltage is applied across the piezoelectric element in the bender assembly causes the attached plastic blades to flap. This principle is illustrated in Fig. 8-13. The highly "focused" air streams are responsible for the unit's exceptional efficiency; the fan moves 20 cubic feet/minute of air while using just 0.36 watts.

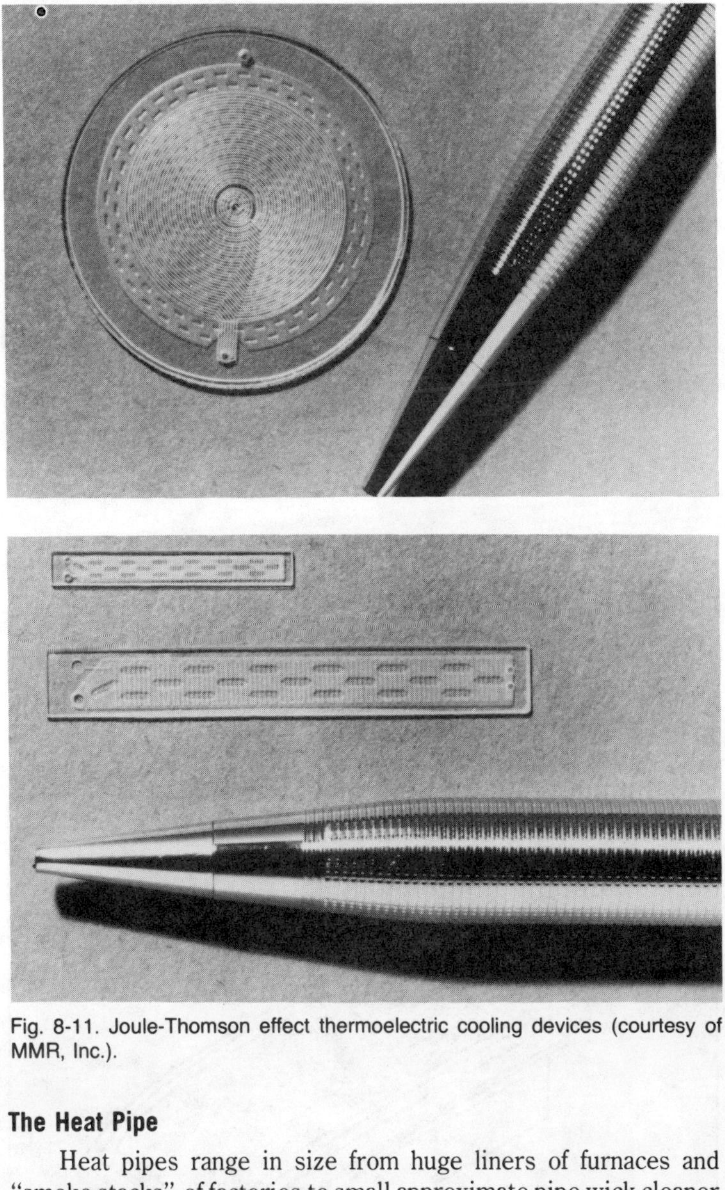

Fig. 8-11. Joule-Thomson effect thermoelectric cooling devices (courtesy of MMR, Inc.).

The Heat Pipe

Heat pipes range in size from huge liners of furnaces and "smoke stacks" of factories to small approximate pipe wick cleaner sized and shaped objects (refer to Fig. 8-14). This highly thermally conductive tube's claim to fame is its unique ability to very equally and uniformly distribute heat and thus equaling out the differences between the hottest and coolest objects (usually ICs and ambient air

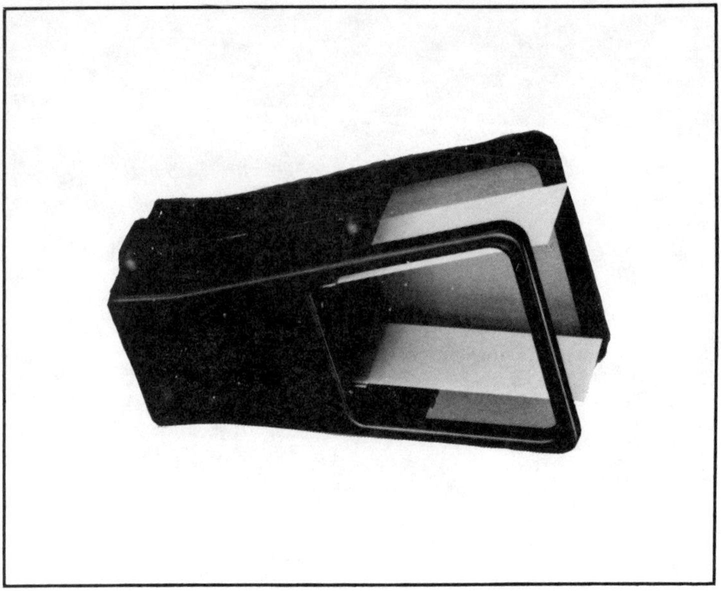

Fig. 8-12. A piezoelectric fan (courtesy of Piezo Electric Products, Inc.).

respectively). These smaller sized heat pipes fit nicely underneath and between the two adjacent rows of IC pins.

Heat pipes are metallic sealed self-contained units (refer to Fig. 8-15) with a thin outer wall or shell of copper which surrounds

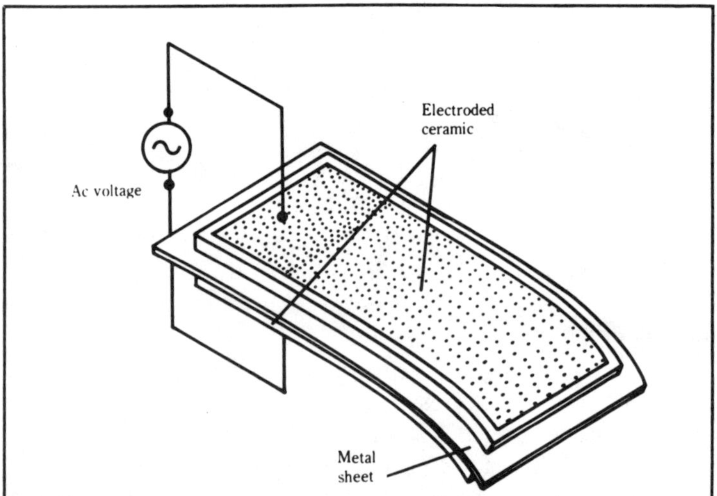

Fig. 8-13. A piezoceramic element, the basis for the piezoelectric fan.

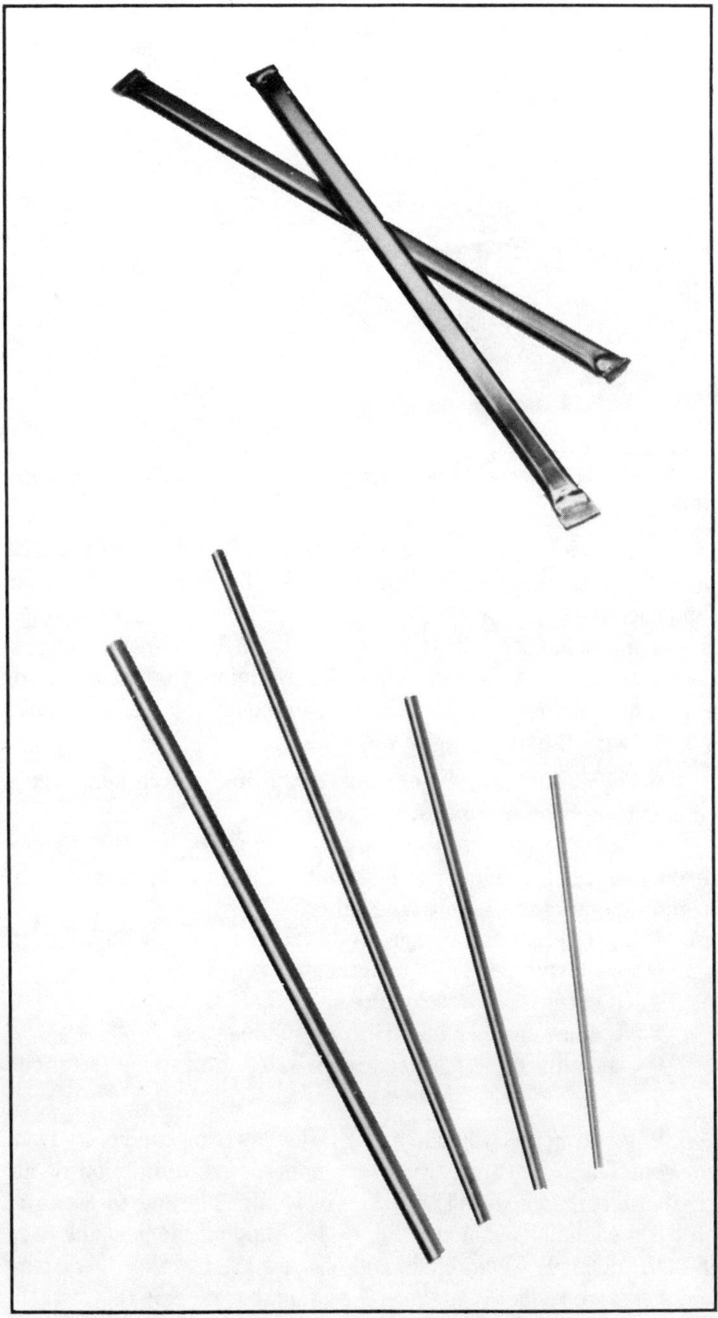

Fig. 8-14. Typical small heat pipes (courtesy of Noren Products, Inc.).

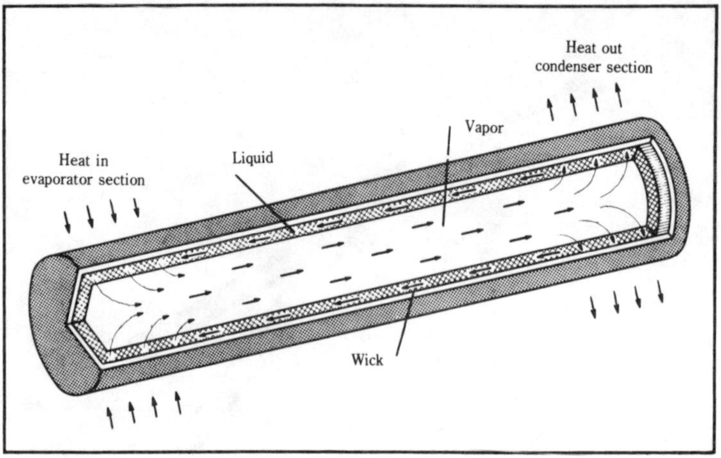

Fig. 8-15. An illustration of the heat pipe's operating principle (courtesy of Noren Products, Inc.).

an internal wick which has been saturated with a working fluid. As one end of the heat pipe experiences heat (the evaporator end), the fluid vaporizes and travels to the other end (the condenser end). As the vapor condenses, heat is given off and the vaporized liquid condenses and travels back, via capillary action, to the evaporator end. The end result is an efficient self-contained heat exchange system with the following advantages:

- The ability to transfer over 1,000 times more heat than a copper rod of the same size and weight.
- Virtually absolutely uniform heat transfer and distribution throughout the heat pipe with less than 0.1° C drop or heat gradient from one end of the pipe to the other.
- Lightweight and compact.
- No moving parts, no maintenance required.
- No external power required.
- No noise—either electrical or audible.
- Capability of being fused to an existing heatsink (refer to Fig. 8-16).

If, for example, you had an LED display composed of a row of adjacent DIPs and there was very poor or no ventilation or air circulation present, as in Fig. 8-17, you wouldn't be able to use a slip on DIP heatsink as shown in Fig. 8-18. Slipping the heatsink over the DIP displays would block and obscure the front of the display and therefore render it useless. A heat pipe is ideal in this application. Often the requirement still exists to vent heat away after it has been transferred and evenly distributed. This is easily done by attaching a radiator, if you will, to the heat pipe, as seen in Fig. 8-19.

TEMPERATURE INDICATORS

After heatsinking and adding either a fan and/or a Peltier or

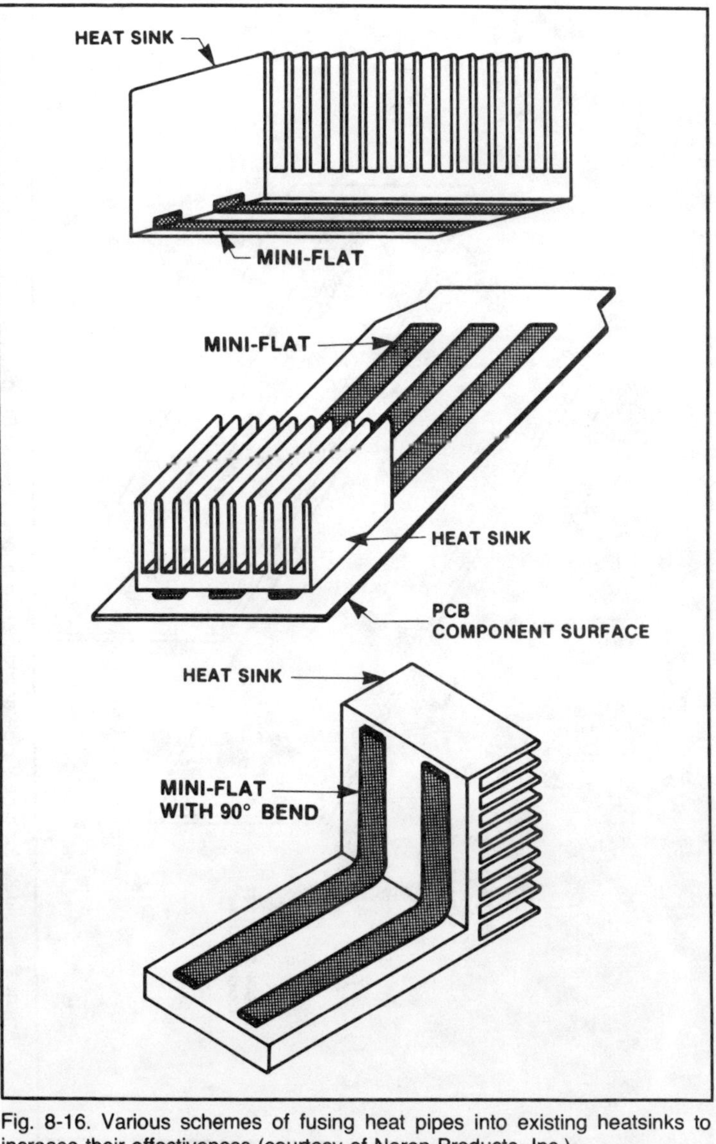

Fig. 8-16. Various schemes of fusing heat pipes into existing heatsinks to increase their effectiveness (courtesy of Noren Products, Inc.).

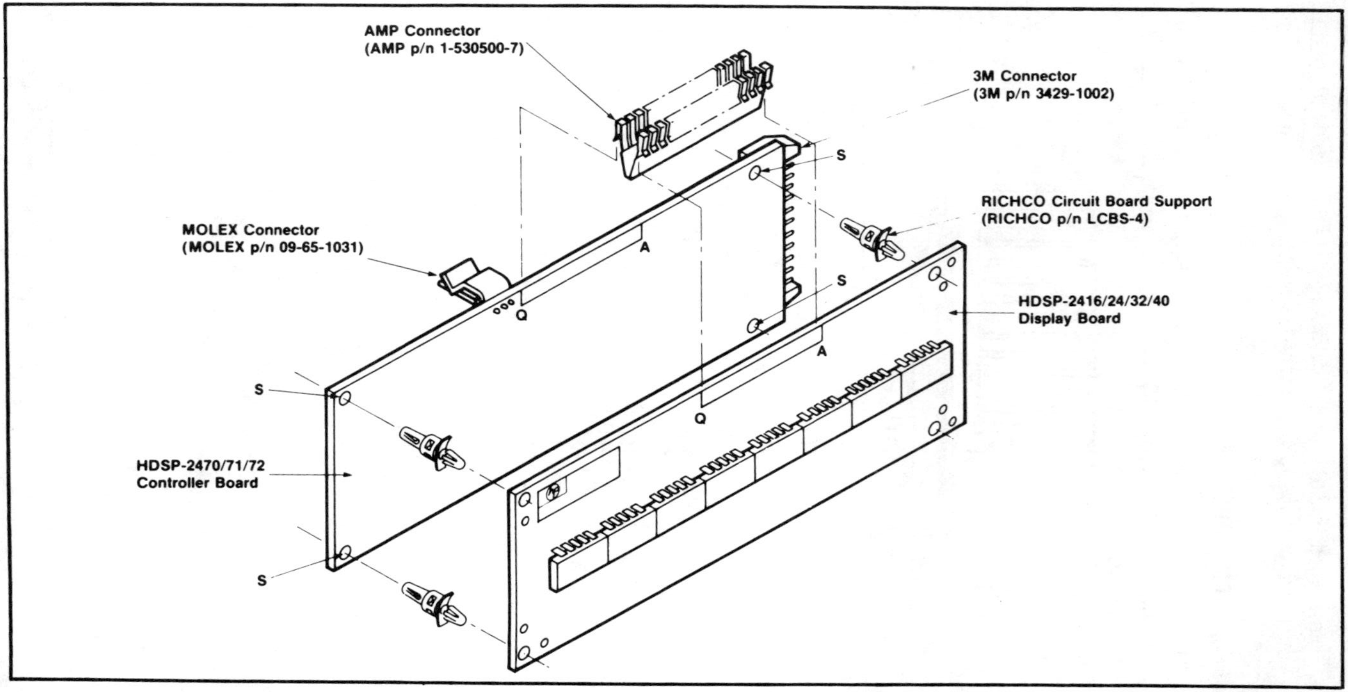

Fig. 8-17. An ideal candidate for a heat pipe application, a poorly vented LED display.

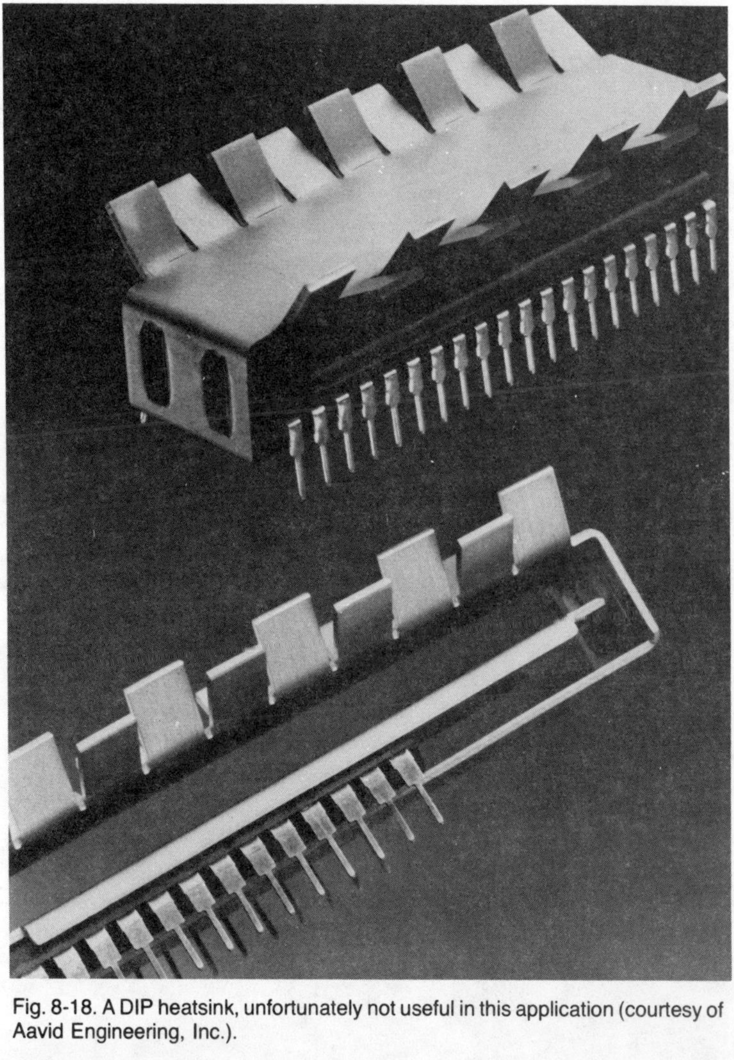

Fig. 8-18. A DIP heatsink, unfortunately not useful in this application (courtesy of Aavid Engineering, Inc.).

Thomson-Joule device, the temperature should be within acceptable limits. Now, let's monitor this temperature. There are a number of companies such as Omega and Telatemp that make crayon appearing objects that rub onto a semiconductor's surface whose temperature is to be monitored (refer to Fig. 8-20). There are also TO-3, TO-66, and DIP stick-on temperature indicators, shown in Fig. 8-21 and a whole host of other products (Fig. 8-22) such as paints and tablets, etc., that are likewise heat sensitive and

Fig. 8-19. A "radiator" attached to a heat pipe for more effective heat radiation (courtesy of Noren Products, Inc.).

exhibit the property of changing colors in direct relationship to the temperature they monitor. As the temperature changes, so does the color of these temperature indicators. These indicators come in two types. The more traditional throw away types change color as temperature rises, but if the temperature falls again, they cannot change back to a color corresponding to this lower temperature. This unfortunate "racket" effect has been overcome by an LCD (liquid crystal display) temperature indicator. These LCD indicators do not use the more traditional nematic crystals but rather use cholesteric crystals which are produced in the cholesterol in lambs wool and in cuttlefish. A thorough explanation of LCDs and the liquid crystal phenomenon may be found in the "LCD Primer"

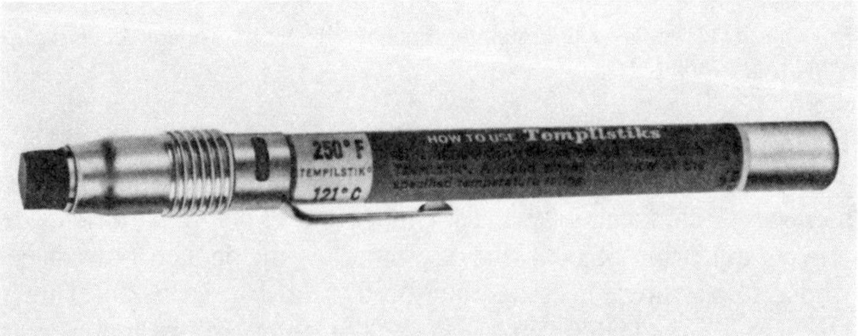

Fig. 8-20. Heat sensitive crayons used to visually detect temperature changes (courtesy of Tempil Division, Big Three Industries, Inc.).

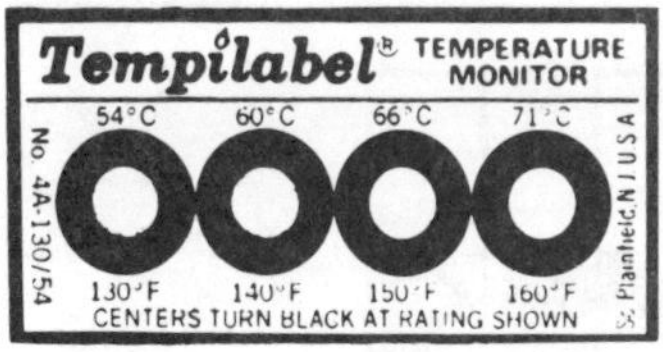

Fig. 8-21. Heat sensitive sticky backed peel off temperature indicators (courtesy of Tempil Division, Big Three Industries, Inc.).

published in the April 1983 *CQ* magazine. These indicators cost about four dollars each, have as many as seven colors, and have 5° C increments as standard. There are larger indicators based on the same principle that can cover whole power supplies or entire printed circuit boards.

Fig. 8-22. A whole product line of visual temperature indicators (courtesy of Omega Engineering, Inc.).

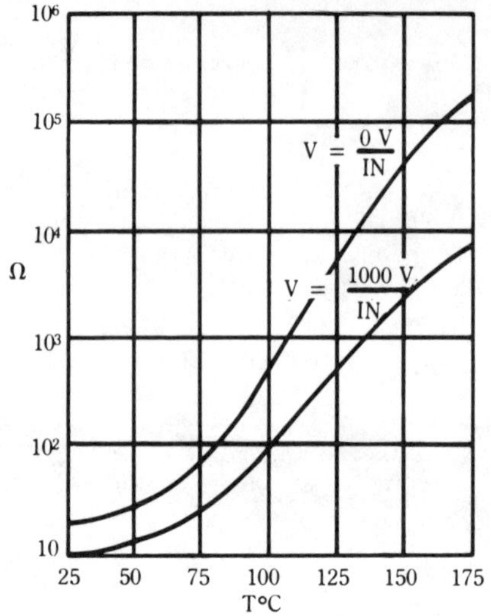

Fig. 8-23. A graph showing how a PTC can overcome the effects of internal device heat generation. This graph demonstrates the sensitivity of resistance of PTC material to applied voltage. In most PTC applications, this voltage sensitivity of resistance or varistor effect is undesirable.

OVERHEATING PROTECTION DEVICES

Overheating can occur and be controlled, but first it has to be sensed. The visual methods just described and shown in Figs. 8-20 through Fig. 8-22 are one form of indicator; however, in unattended applications, there are electrical parameters as well as heat that must be sensed and controlled. In a typical power supply, there is a fuse at the transformer, but other protection can be given to individual component parts within the system.

A PTC (positive temperature coefficient) thermistor can help. These devices are very inexpensive, typically one dollar, and are two-lead devices that sense ambient heat by having the body or encapsulating cover experience heating, causing its resistance to rise by an order of magnitude (a factor of ten) for each 10° C rise in temperature above its trip point. This trip point is different per unit and can be specified, but a typical trip point is about 100° C. Once this elevated temperature is reached though, automatic reset is not possible by merely removing power to the PTC thermistor. An actual cooling of the PTC thermistor to below its trip temperature must first occur before resetting is possible. The internal heating effect of the thermistor itself is inadequate to cause it to reach the trip temperature. This allows internal heating to have only a very small effect on the resistance of this heat sensitive resistor. It also allows the surface that is producing heat, usually flush with the PTC thermistor, to control the device's resistance. Figure 8-23 is a graph showing how this internal heating effect is minimized by a PTC.

Appendix

National Semiconductor

LM723/LM723C Voltage Regulator

General Description

The LM723/LM723C is a voltage regulator designed primarily for series regulator applications. By itself, it will supply output currents up to 150 mA; but external transistors can be added to provide any desired load current. The circuit features extremely low standby current drain, and provision is made for either linear or foldback current limiting. Important characteristics are:

- 150 mA output current without external pass transistor
- Output currents in excess of 10A possible by adding external transistors
- Input voltage 40V max
- Output voltage adjustable from 2V to 37V
- Can be used as either a linear or a switching regulator.

The LM723/LM723C is also useful in a wide range of other applications such as a shunt regulator, a current regulator or a temperature controller.

The LM723C is identical to the LM723 except that the LM723C has its performance guaranteed over a 0°C to 70°C temperature range, instead of −55°C to +125°C.

Schematic and Connection Diagrams *

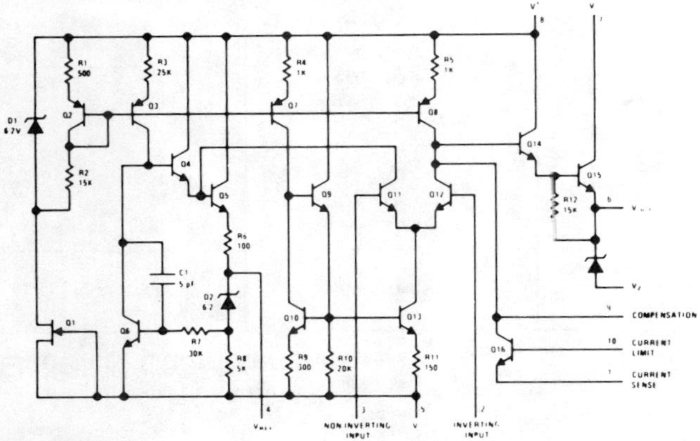

Dual-In-Line Package

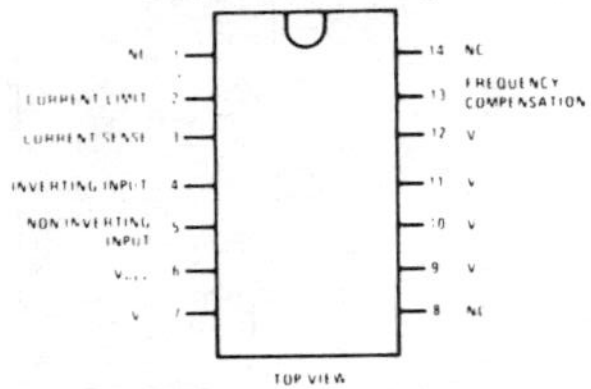

Order Number LM723CN
See NS Package N14A
Order Number LM723J or LM723CJ
See NS Package J14A

Metal Can Package

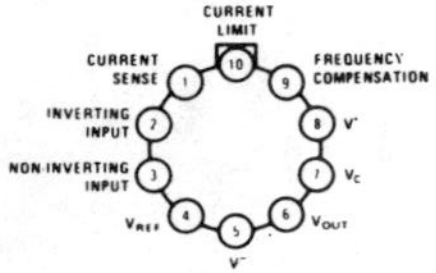

Note: Pin 5 connected to case.

TOP VIEW

Order Number LM723H or LM723CH
See NS Package H10C

Equivalent Circuit *

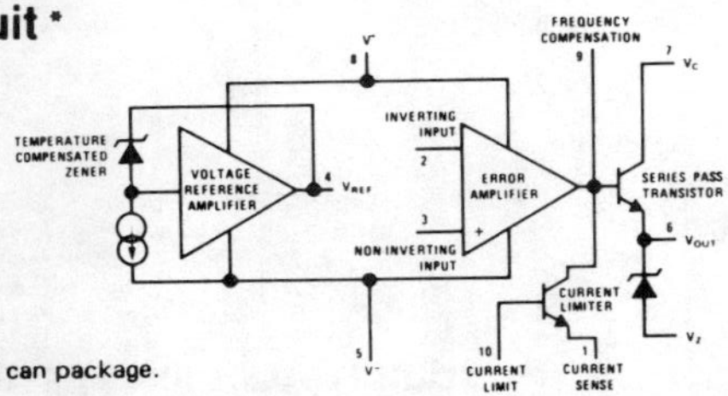

*Pin numbers refer to metal can package.

Absolute Maximum Ratings

Pulse Voltage from V^+ to V^- (50 ms)	50V
Continuous Voltage from V^+ to V^-	40V
Input-Output Voltage Differential	40V
Maximum Amplifier Input Voltage (Either Input)	7.5V
Maximum Amplifier Input Voltage (Differential)	5V
Current from V_Z	25 mA
Current from V_{REF}	15 mA
Internal Power Dissipation Metal Can (Note 1)	800 mW
Cavity DIP (Note 1)	900 mW
Molded DIP (Note 1)	660 mW
Operating Temperature Range LM723	-55°C to +125°C
LM723C	0°C to +70°C
Storage Temperature Range Metal Can	-65°C to +150°C
DIP	-55°C to +125°C
Lead Temperature (Soldering, 10 sec)	300°C

Electrical Characteristics (Note 2)

PARAMETER	CONDITIONS	LM723			LM723C			UNITS
		MIN	TYP	MAX	MIN	TYP	MAX	
Line Regulation	$V_{IN} = 12V$ to $V_{IN} = 15V$		.01	0.1		.01	0.1	% V_{OUT}
	$-55°C < T_A < +125°C$			0.3				% V_{OUT}
	$0°C < T_A < +70°C$						0.3	% V_{OUT}
	$V_{IN} = 12V$ to $V_{IN} = 40V$		.02	0.2		0.1	0.5	% V_{OUT}
Load Regulation	$I_L = 1$ mA to $I_L = 50$ mA		.03	0.15		.03	0.2	% V_{OUT}
	$-55°C < T_A < +125°C$			0.6				% V_{OUT}
	$0°C < T_A <= +70°C$						0.6	% V_{OUT}
Ripple Rejection	f = 50 Hz to 10 kHz, $C_{REF} = 0$		74			74		dB
	f = 50 Hz to 10 kHz, $C_{REF} = 5\,\mu F$		86			86		dB

Average Temperature Coefficient of Output Voltage	$-55°C < T_A < +125°C$		.002	.015				%/°C
	$0°C \leq T_A < +70°C$					.003	.015	%/°C
Short Circuit Current Limit	$R_{SC} = 10\Omega$, $V_{OUT} = 0$		65			65		mA
Reference Voltage		6.95	7.15	7.35	6.80	7.15	7.50	V
Output Noise Voltage	BW = 100 Hz to 10 kHz, $C_{REF} = 0$		20			20		μVrms
	BW = 100 Hz to 10 kHz, $C_{REF} = 5\,\mu F$		2.5			2.5		μVrms
Long Term Stability			0.1			0.1		%/1000 hrs
Standby Current Drain	$I_L = 0$, $V_{IN} = 30V$		1.3	3.5		1.3	4.0	mA
Input Voltage Range		9.5		40	9.5		40	V
Output Voltage Range		2.0		37	2.0		37	V
Input-Output Voltage Differential		3.0		38	3.0		38	V

Note 1: See derating curves for maximum power rating above 25°C.

Note 2: Unless otherwise specified, $T_A = 25°C$, $V_{IN} = V^+ = V_C = 12V$, $V^- = 0$, $V_{OUT} = 5V$, $I_L = 1$ mA, $R_{SC} = 0$, $C_1 = 100$ pF, $C_{REF} = 0$ and divider impedance as seen by error amplifier $\leq 10\ k\Omega$ connected as shown in Figure 1. Line and load regulation specifications are given for the condition of constant chip temperature. Temperature drifts must be taken into account separately for high dissipation conditions.

Note 3: L_1 is 40 turns of No. 20 enameled copper wire wound on Ferroxcube P36/22-3B7 pot core or equivalent with 0.009 in. air gap.

Note 4: Figures in parentheses may be used if R1/R2 divider is placed on opposite input of error amp.

Note 5: Replace R1/R2 in figures with divider shown in Figure 13.

Note 6: V^+ must be connected to a +3V or greater supply.

Note 7: For metal can applications where V_Z is required, an external 6.2 volt zener diode should be connected in series with V_{OUT}.

Maximum Power Ratings

LM723

Power Dissipation vs Ambient Temperature

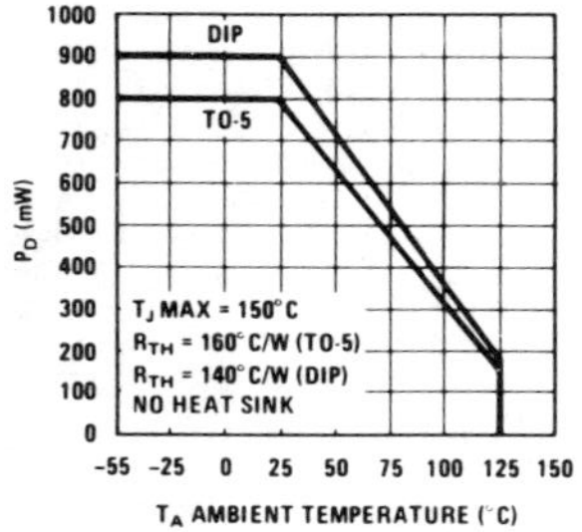

LM723C

Power Dissipation vs Ambient Temperature

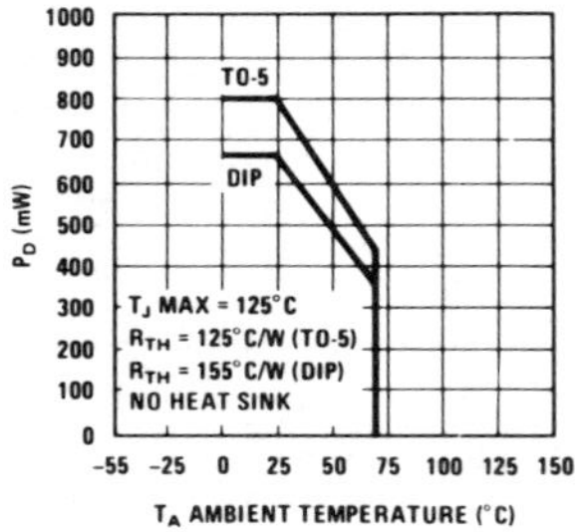

Typical Performance Characteristics

Load Regulation Characteristics with Current Limiting

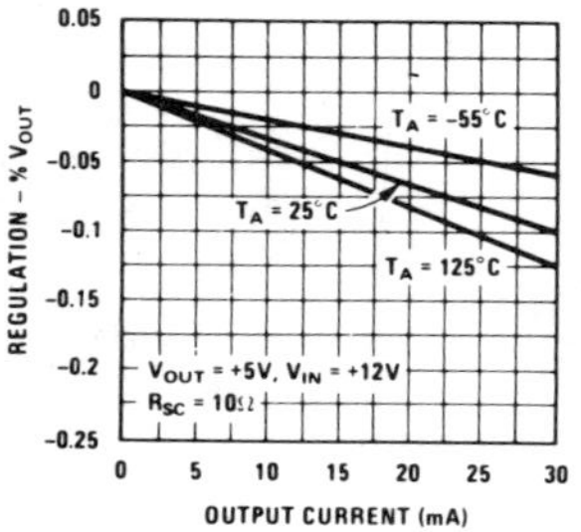

Load Regulation Characteristics with Current Limiting

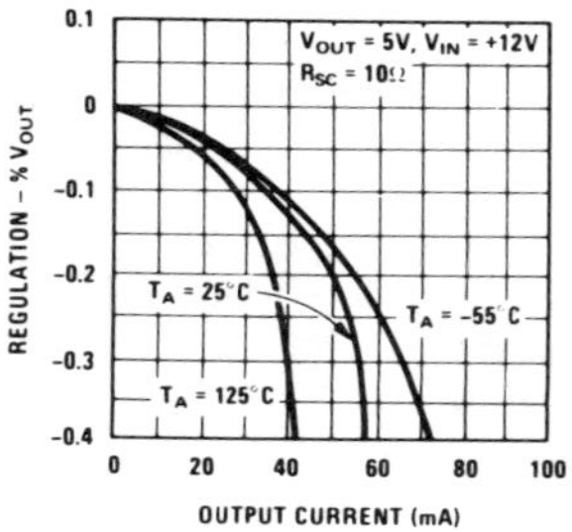

Load & Line Regulation vs Input-Output Voltage Differential

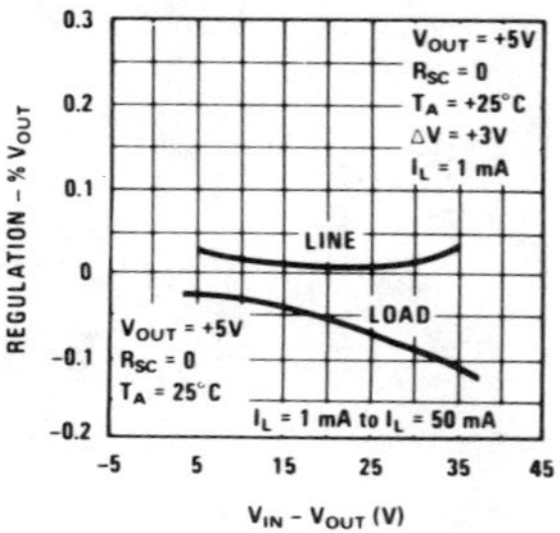

Current Limiting Characteristics

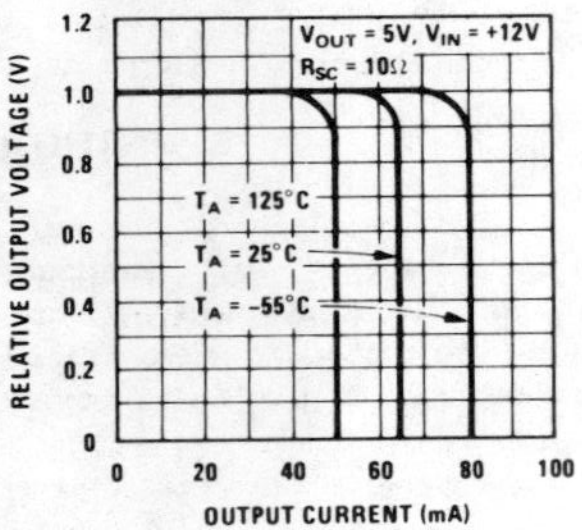

Current Limiting Characteristics vs Junction Temperature

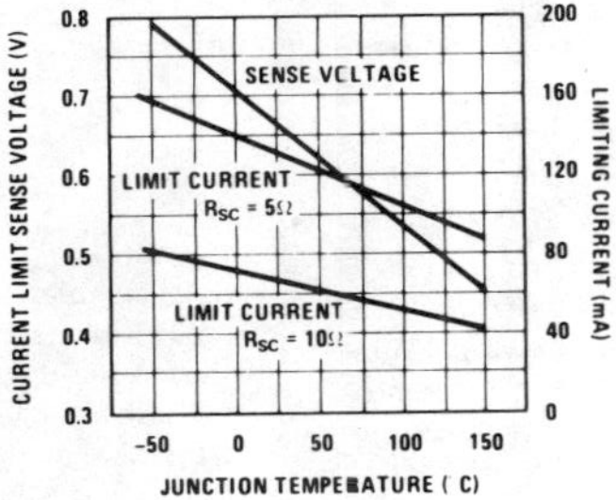

Standby Current Drain vs Input Voltage

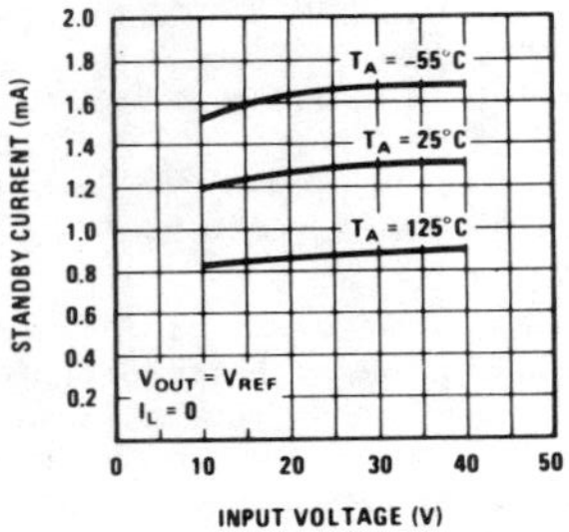

Line Transient Response

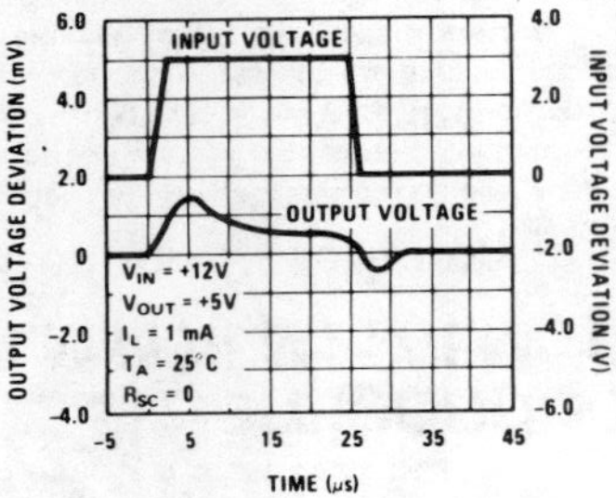

Load Transient Response

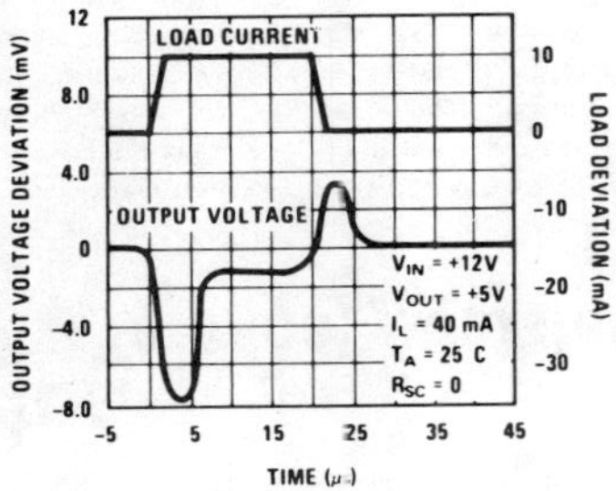

Output Impedance vs Frequency

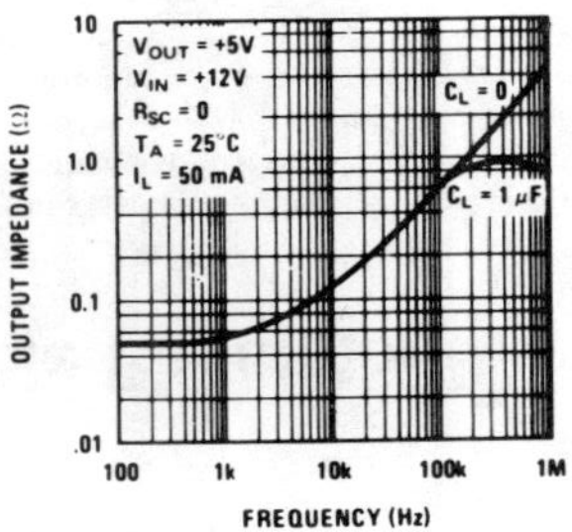

National Semiconductor

LM117/LM217/LM317 3-Terminal Adjustable Regulator

General Description

The LM117/LM217/LM317 are adjustable 3-terminal positive voltage regulators capable of supplying in excess of 1.5A over a 1.2V to 37V output range. They are exceptionally easy to use and require only two external resistors to set the output voltage. Further, both line and load regulation are better than standard fixed regulators. Also, the LM117 is packaged in standard transistor packages which are easily mounted and handled.

In addition to higher performance than fixed regulators, the LM117 series offers full overload protection available only in IC's. Included on the chip are current limit, thermal overload protection and safe area protection. All overload protection circuitry remains fully functional even if the adjustment terminal is disconnected.

Features

- Adjustable output down to 1.2V
- Guaranteed 1.5A output current
- Line regulation typically 0.01%/V
- Load regulation typically 0.1%
- Current limit constant with temperature
- **100% electrical burn-in**

Besides replacing fixed regulators, the LM117 is useful in a wide variety of other applications. Since the regulator is "floating" and sees only the input-to-output differential voltage, supplies of several hundred volts can be regulated as long as the maximum input to output differential is not exceeded.

Also, it makes an especially simple adjustable switching regulator, a programmable output regulator, or by connecting a fixed resistor between the adjustment and output, the LM117 can be used as a precision current regulator. Supplies with electronic shutdown can be achieved by clamping the adjustment terminal to ground which programs the output to 1.2V where most loads draw little current.

The LM117K, LM217K and LM317K are packaged in standard TO-3 transistor packages while the LM117H, LM217H and LM317H are packaged in a solid Kovar base TO-5 transistor package. The LM117 is rated for operation from −55°C to +150°C, the LM217 from −25°C to +150°C and the LM317 from 0°C to +125°C. The LM317T and LM317MP, rated for operation over a 0°C to +125°C range, are available in a TO-220 plastic package and a TO-202 package, respectively.

For applications requiring greater output current in

- Eliminates the need to stock many voltages
- Standard 3-lead transistor package
- 80 dB ripple rejection

Normally, no capacitors are needed unless the device is situated far from the input filter capacitors in which case an input bypass is needed. An optional output capacitor can be added to improve transient response. The adjustment terminal can be bypassed to achieve very high ripple rejections ratios which are difficult to achieve with standard 3-terminal regulators.

excess of 3A and 5A, see LM150 series and LM138 series data sheets, respectively. For the negative complement, see LM137 series data sheet.

LM117 Series Packages and Power Capability

DEVICE	PACKAGE	RATED POWER DISSIPATION	DESIGN LOAD CURRENT
LM117	TO-3	20W	1.5A
LM217 LM317	TO-5	2W	0.5A
LM317T	TO-220	15W	1.5A
LM317M	TO-202	7.5W	0.5A

Typical Applications

1.2V–25V Adjustable Regulator

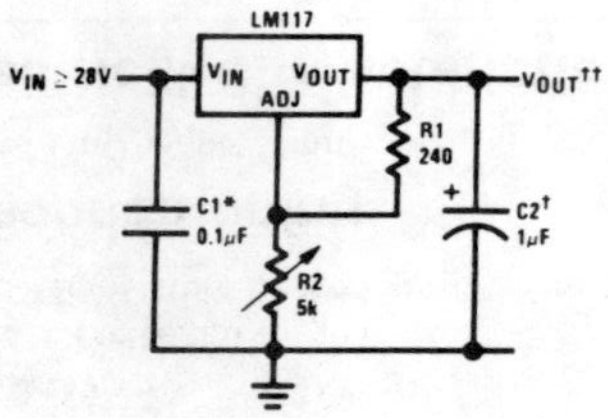

†Optional—improves transient response

*Needed if device is far from filter capacitors

$$^{\dagger\dagger}V_{OUT} = 1.25V \left(1 + \frac{R2}{R1}\right)$$

Digitally Selected Outputs

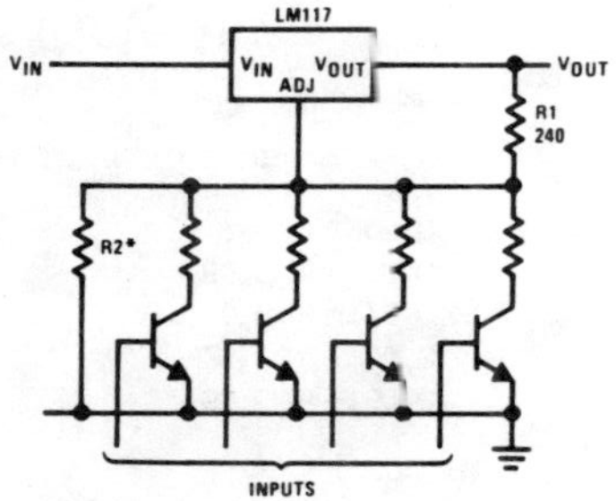

*Sets maximum V_{OUT}

5V Logic Regulator with Electronic Shutdown*

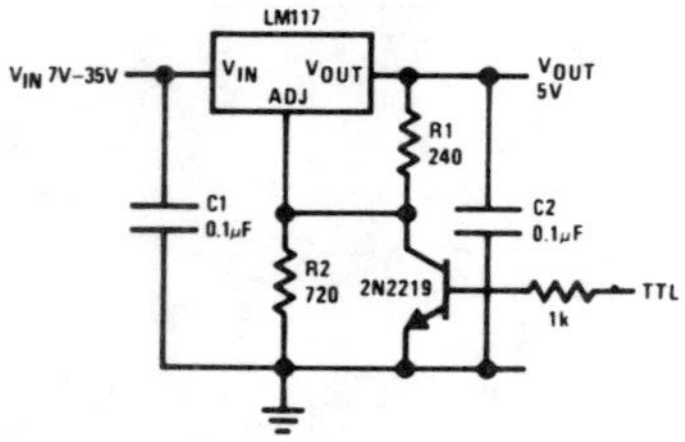

*Min output ≈ 1.2V

Absolute Maximum Ratings

Power Dissipation	Internally limited
Input–Output Voltage Differential	40V
Operating Junction Temperature Range	
LM117	−55°C to +150°C
LM217	−25°C to +150°C
LM317	0°C to +125°C
Storage Temperature	−65°C to +150°C
Lead Temperature (Soldering, 10 seconds)	300°C

Preconditioning

Burn-In in Thermal Limit **100% All Devices**

Electrical Characteristics (Note 1)

PARAMETER	CONDITIONS	LM117/217			LM317			UNITS
		MIN	TYP	MAX	MIN	TYP	MAX	
Line Regulation	$T_A = 25°C$, $3V \leq V_{IN} - V_{OUT} \leq 40V$ (Note 2)		0.01	0.02		0.01	0.04	%/V
Load Regulation	$T_A = 25°C$, $10\ mA \leq I_{OUT} \leq I_{MAX}$							
	$V_{OUT} \leq 5V$, (Note 2)		5	15		5	25	mV
	$V_{OUT} \geq 5V$, (Note 2)		0.1	0.3		0.1	0.5	%
Thermal Regulation	$T_A = 25°C$, 20 ms Pulse		0.03	0.07		0.04	0.07	%/W
Adjustment Pin Current			50	100		50	100	μA
Adjustment Pin Current Change	$10\ mA \leq I_L \leq I_{MAX}$ $2.5V \leq (V_{IN}-V_{OUT}) \leq 40V$		0.2	5		0.2	5	μA
Reference Voltage	$3 \leq (V_{IN}-V_{OUT}) \leq 40V$, (Note 3) $10\ mA \leq I_{OUT} \leq I_{MAX}$, $P \leq P_{MAX}$	1.20	1.25	1.30	1.20	1.25	1.30	V

Line Regulation	$3V \leq V_{IN} - V_{OUT} \leq 40V$, (Note 2)		0.02	0.05		0.02	0.07	%/V
Load Regulation	$10\ mA \leq I_{OUT} \leq I_{MAX}$, (Note 2)							
	$V_{OUT} \leq 5V$		20	50		20	70	mV
	$V_{OUT} \geq 5V$		0.3	1		0.3	1.5	%
Temperature Stability	$T_{MIN} \leq T_j \leq T_{MAX}$		1			1		%
Minimum Load Current	$V_{IN}-V_{OUT} = 40V$		3.5	5		3.5	10	mA
Current Limit	$V_{IN}-V_{OUT} \leq 15V$							
	K and T Package	1.5	2.2		1.5	2.2		A
	H and P Package	0.5	0.8		0.5	0.8		A
	$V_{IN}-V_{OUT} = 40V$							
	K and T Package		0.4			0.4		A
	H and P Package		0.07			0.07		A
RMS Output Noise, % of V_{OUT}	$T_A = 25°C$, $10\ Hz \leq f \leq 10\ kHz$		0.003			0.003		%
Ripple Rejection Ratio	$V_{OUT} = 10V$, $f = 120\ Hz$		65			65		dB
	$C_{ADJ} = 10\mu F$	66	80		66	80		dB
Long-Term Stability	$T_A = 125°C$		0.3	1		0.3	1	%
Thermal Resistance, Junction to Case	H Package		12	15		12	15	°C/W
	K Package		2.3	3		2.3	3	°C/W
	T Package					4		°C/W
	P Package					12		°C/W

Note 1: Unless otherwise specified, these specifications apply: $-55°C \leq T_j \leq +150°C$ for the LM117, $-25°C \leq T_j \leq +150°C$ for the LM217 and $0°C \leq T_j \leq +125°C$ for the LM317; $V_{IN}-V_{OUT} = 5V$ and $I_{OUT} = 0.1A$ for the TO-5 and TO-202 packages and $I_{OUT} = 0.5A$ for the TO-3 package and TO-220 package. Although power dissipation is internally limited, these specifications are applicable for power dissipations of 2W for the TO-5 and TO-202 and 20W for the TO-3 and TO-220. I_{MAX} is 1.5A for the TO-3 and TO-220 package and 0.5A for the TO-5 and TO-202 package.

Note 2: Regulation is measured at constant junction temperature, using pulse testing with a low duty cycle. Changes in output voltage due to heating effects are covered under the specification for thermal regulation.

Note 3: Selected devices with tightend tolerance reference voltage available.

Typical Performance Characteristics (K and T Packages)

Load Regulation

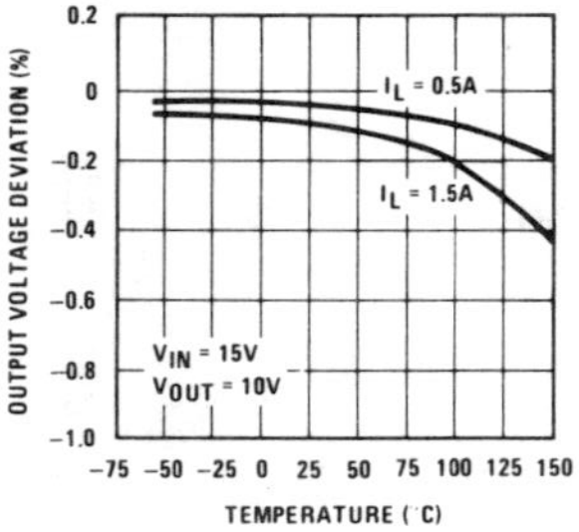

Current Limit

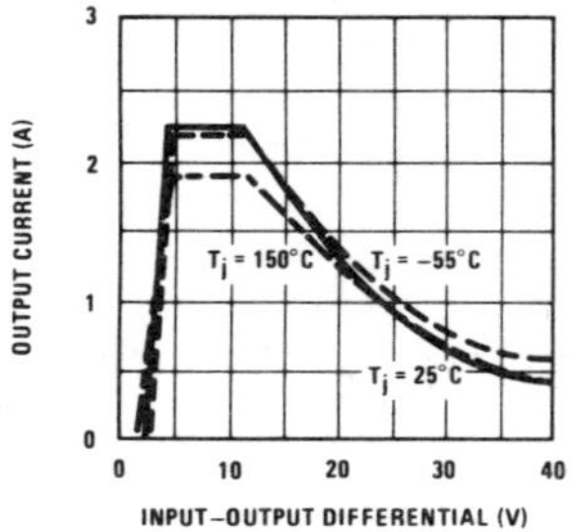

Adjustment Current

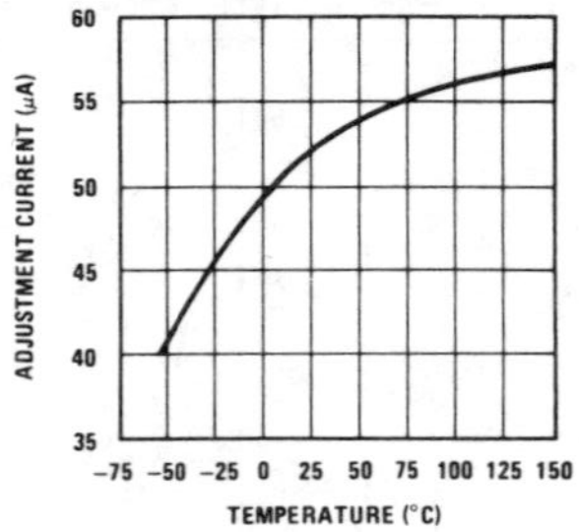

Dropout Voltage

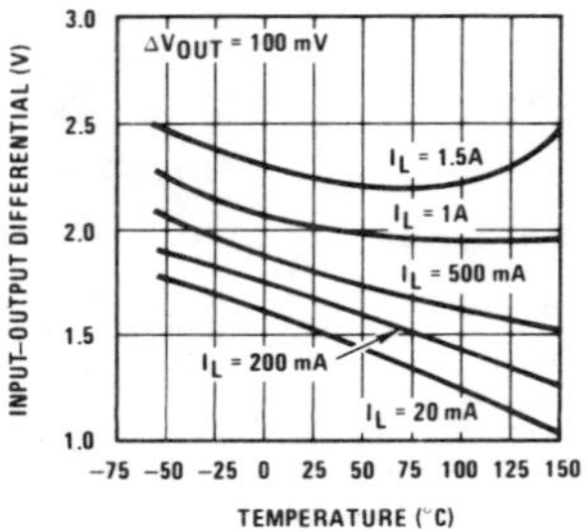

Temperature Stability

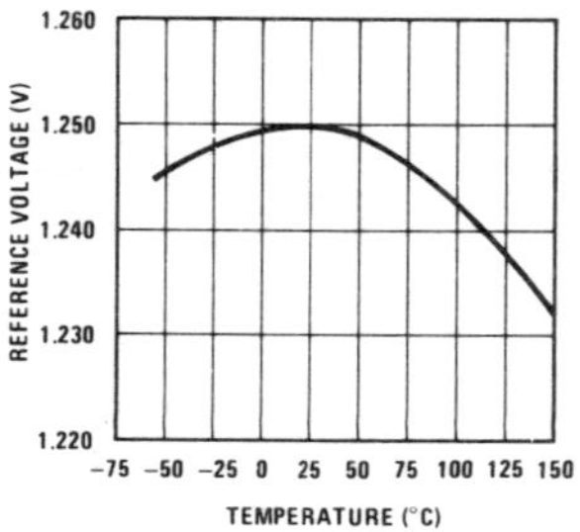

Minimum Operating Current

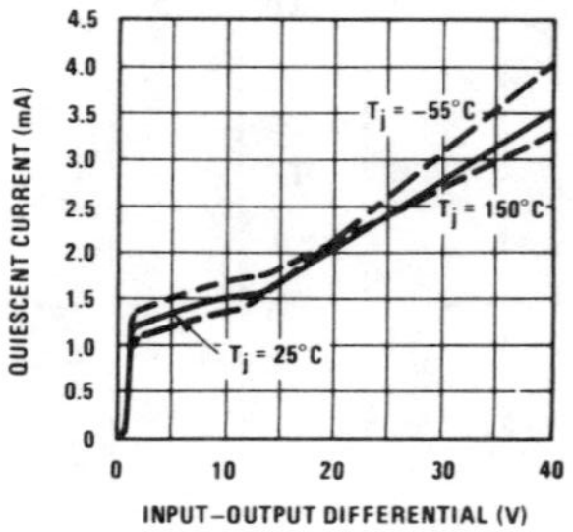

Ripple Rejection

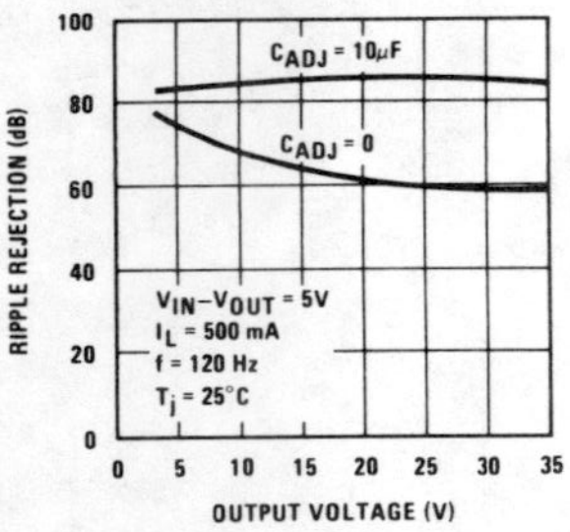

Ripple Rejection

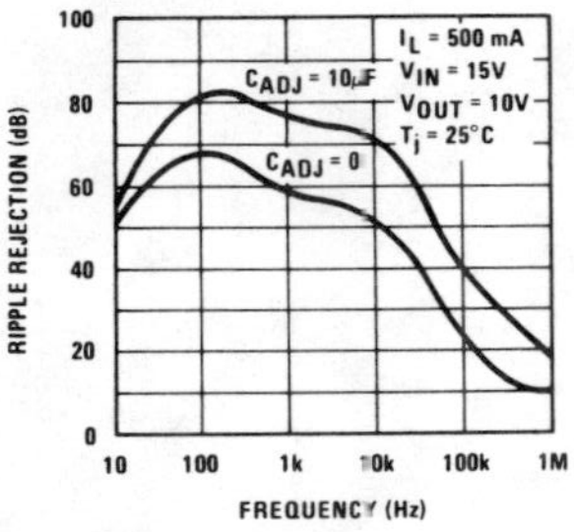

Ripple Rejection

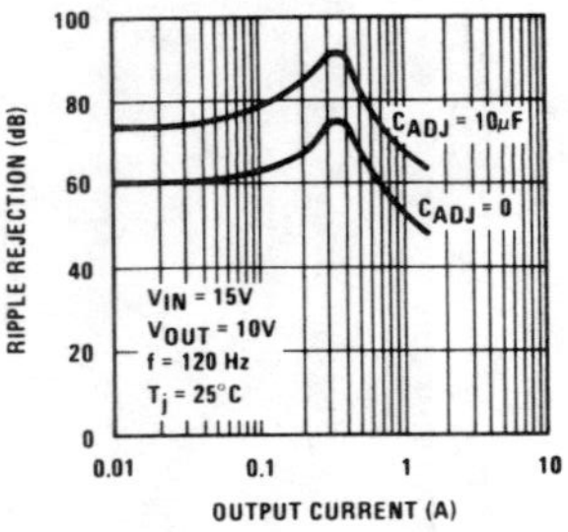

Output Impedance

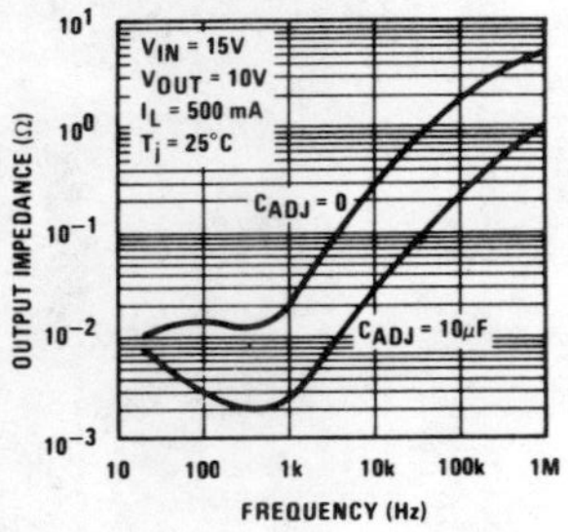

Line Transient Response

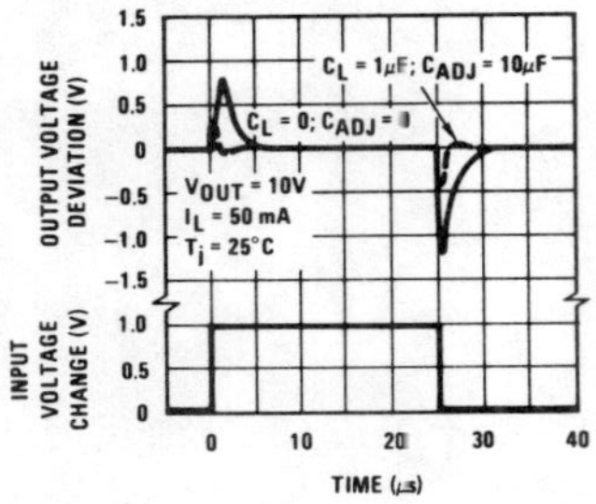

Load Transient Response

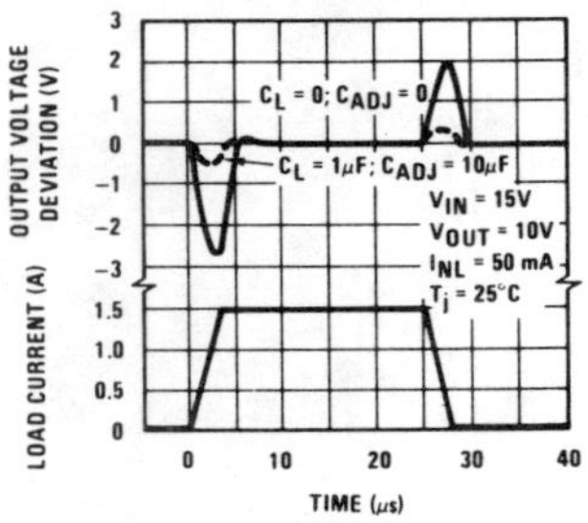

Bibliography

VOLTAGE REFERENCES AND ZENERS

Brokaw, Paul. "Simplified Voltage-Splitting Regulator Neatly Divides Single Power Source", *Electronic Design*, October 11, 1979, p. 166.

Dobkin, Robert C. "IC Voltage References," *Electronic Products Magazine*, May, 1978, p. 89.

Downs, Stuart G. "Stable Voltage Reference Uses Microamps," *EDN*, August 5, 1980, p. 129.

Freeman, Wes, and Erdi, George. "Stacked Voltage References Improve Supply's Regulation," *Electronics*, September 22, 1981, p. 132.

Hafer, C.R. Owen, Neil. "Low-Power Precision Reference Reduces Long Term Drift," *EDN*, May 5, 1978, p. 99.

Hanna, Kim. "Single Control Adjusts Voltage Reference," *EDN*, June 5, 1976, p. 117.

Jung, Walter G. "Precision Reference Source Features Minimum Errors," *EDN*, August 5, 1976, p. 80.

———. "Simple Regulator Circuits Provide Tracking Bipolar Reference Voltages," *Electronic Design*, July 5, 1976, p. 62.

———. "Positive Reference-Voltage IC is Flipped Negative by Adding a Single Component," *Electronic Design*, February 15, 1978, p. 98.

———. "Voltage References Determine Accuracy," *Electronic Design*, August 2, 1977, p. 84.

———. "Precision Reference Source Eliminates Power-Supply

Regulation Errors," *Electronic Design*, September 1, 1978, p. 108.

———. "Programmable Voltage Reference is Stable, Yet Simple," *EDN*, November 5, 1975, p. 98.

Koep, Ken. "If You Want a High-Accuracy Zener, Specify Time Stability, Not Temperature Coefficient,"*Electronic Design*, May 10, 1978, p. 132.

McDermott, Jim. "Ultraprecision IC Voltage References Serve Varied Circuit Needs," *EDN*, April 1, 1981, p. 61.

Nelson, Carl. "Precision Reference uses only 10 mA", *EDN*, May 20, 1979, p. 189.

Normet, Henno. "High-Accuracy Calibrator Uses Bandgap Voltage Reference," *Electronics*, June 8, 1978, p. 118.

Panck, Kenneth E. 'Age Reference Zeners to Achieve Stable Long-Term Performance," *Electronic Design*, September 13, 1976, p. 120.

Pouliot, Don and Calkins, Bob. "Build Precision Reference Sources," *Electronic Design*, July 5, 1976, p. 56.

Sass, Forrest and Welling, Brent. 'Use an IC Voltage Reference to Control Crystal-Oven Temperature Accurately," *Electronic Design*, September 13, 1976, p. 124.

Sevastopoulos, Nello and Moyer, Jim. "Micropower Reference Stays Stable," *EDN* September 5, 1978, p. 158.

CONSTANT-CURRENT SOURCES/SINKS

Baily, John T. "Nanoampere Current Source is Adjustable, Stable, Inexpensive," *Electronic Design*, January 22, 1981, p. 134.

Bober, Robert. "Dynamic Load Requires no Additional Power Supplies," *EDN*, March 20, 1975, p. 76.

Gardner, Mark R. "Build a Constant-Current Source or Sink with an Unused Comparator of a Quad Package," *Electronic Design*, February 1, 1978, p. 86.

Jung, Walter G. "Current Inverters Make Current Sources out of Monolithic—IC D/A Sink Outputs," *Electronic Design*, January 18, 1978, p. 108.

Lefferts, Peter. "Variable Slope Current Source Starts at 2.5 V", *EDN*, November 5, 1975, p. 100.

Mundi, William J. "Milliampere Current Source is Voltage Controlled,"*Electronics*, April 7, 1981, p. 134.

Nelson, Carl. "Improve Current Source Performance,"*EDN*, February 20, 1979, p. 157.

Nussbaum, Stephen H. "Current Source for I^2L Saves Energy," *Electronics*, August 16, 1979, p. 115.

Roberts, John A. "Stable Constant-Voltage Supply Uses Current Source to Establish Reference," *Electronic Design*, November 22, 1976, p. 176.

SWITCHING REGULATOR ARTICLES

Apfel, Russell J. "Regulators to Switch or Not?" *Electronic Products Magazine*, August, 1977, p. 39.

Burens, James H. "Switching Power Supplies: Specification Criteria," *Computer Design*, March, 1977, p. 91.

Chambers, Derek and Wang, Dee. "Assessing Switcher Performance Calls for Fresh Ways and Means," *EDN*, April 20, 1978, p. 111.

Chetty, P. R. K. "DC-DC Converter Maintains High-Efficiency," *Electronics*, January 3, 1980, p. 159.

Coon, Art. "IC Voltage Regulators," *Electronic Products Magazine*, December, 1980, p. 51.

Cowett, Philip M. "Switching Supply Converts −60 V to +5 and ±6.3 V with 83% Efficiency," *Electronic Design*, January 18, 1978, p. 106.

Faulkenberry, Luces M. "Low-Cost Timers Govern Switched-Mode Regulator," *Electronics*, December 18, 1980, p. 97.

Frostholm, Robert C. "One-Chip Controller Simplifies Switched-Mode Supplies," *Electronics*, June 21, 1979, p. 14.

———. "Versatile Switching Regulator," *Radio Electronics*, February, 1980, p. 56.

Fung, Henry and Rizzo, John. "Regulator Boosts Supply Voltage for Programming EE-PROMs," *Electronics*, July 14, 1981, p. 124.

Gross, Tom. "A Little Understanding Improves Switching-Inductor Designs," *EDN*, June 20, 1979, p. 129.

Grossman, Morris. "Focus on Switching Power Supplies: Advances Fuel High-Growth Rate," *Electronic Design*, March 5, 1981, p. 147.

Haver, Bob. "Current Capacity of Three-Terminal Regulator Expands Five-fold," *Electronic Design*, December 20, 1979, p. 94.

Hirschberg, Walter J. "Which Multiple-Output Technique Should Your Switching Supply Use? *EDN*, March 5, 1978, p. 91.

Hnatek, Eugene A. "Build DC/DC Converters That Work," *Electronic Design*, May 24, 1975 p. 82.

Hopkins, Tom. "Constant Frequency Pulse-Width Modulator Built with One-Chip Op-Amp and Comparator IC," *Electronic Design*, April 12, 1978, p. 110.

Hudor, Andrew M. "Power Converter Uses Low-Cost Audio Transformer," *EDN*, April 20, 1977, p. 139.

Jones, Dave. "Fluorescent Lamp Lights with Low-Voltage Batteries and a Switching Regulator IC," *Electronic Design*, December 6, 1978, p. 114.

Korba, Larry. "Integrator Improves 555 Pulse-Width Modulator," *Electronics*, April 7, 1981, p. 137.

Kulturel, Sadeddin. "Switching Converter Raises Linear Regulator's Efficiency," *Electronics*, June 21, 1979, p. 126.

Lange, Jim. "Tapped Inductor Improves Efficiency for Switching Regulator," *Electronic Design*, August 2, 1979, p. 114.

Levy, Aron. "Head off DC/DC-Converter Problems," *Electronic Design*, August 16, 1976, p. 72.

McLyman, W. T. "Feedback Arrangement for Regenerative Switches," *NASA Tech Briefs*, Fall, 1976, p. 324.

Montgomery, Ray. "Resolving the Problems of Switching-Mode Power-Supplies with the ZN1066," *Ferranti Semiconductor*, Copyright 1981.

———. "Control IC, Circuit Design Grants Long Life to Switches," *Electronic Design*, August 2, 1980, p. 93.

Ohr, Stephan. "IC Regulators Take More Control of Supplies," *Electronic Design*, August 2, 1980, p. 42.

O'Neal, J. H. "Use a Pulse-Width-Modulator Switcher," *Electronic Design*, April 12, 1977, p. 110.

Parkinson, Joseph. "Proper Filter Choice Eliminates Oscillations in DC-DC Converters," *Electronics*, June 8, 1978, p. 146.

Pivit, Erich and Saxarra, J. "Upgrade Your Switchers Analytically," *Electronic Design*, May 10, 1978, p. 108.

Schneider, Hal. *Vari-Drive the Big Difference in Switchers*, Scientific Sales Corporation, 1981, p. 14.

Severns, Rudy. *The Design of Switchmode Converters Above 100 kHz*, Intersil Application Bulletin A034, Copyright 1980.

Shecora, Joseph J. "Regulator Switches Quickly," *EDN*, January 20, 1979, p. 108.

Sheehan, Dan. "Designing a Regulator's LC Input Filter: 'Ripple' Method Prevents Oscillation Woes," *Electronic Design*, August 2, 1979, p. 102.

Spencer, John. "Monolithic Switching Regulators—They Fit Today's Power-Supply Needs," *EDN*, September 5, 1977, p. 117.

Teja, Edward R. "Switching Power-Supplies," *EDN*, February 4, 1981, p. 95.

Thoen, Bradford K. "Solid-State Reference Potential Translator Replaces Current-Sense Transformer in Switching-Mode Power-Supply", *Electronic Design*, March 5, 1981, p. 139.

Thompson, Jeff. "A Comparison of Switching Regulators," *Fairchild Semiconductor Progress*, June/July, 1981.

Tkacenko, Nikola. "Transformer Increases DC-DC Converter Efficiency to 80%," *EDN*, May 5, 1976, p. 110.

van der Poel, Jan. "Pick the Right DC/DC Converter," *Electronic Design*, June 7, 1978, p. 104.

Williams, Jim. "Design DC-DC Converters to Catch Noise at the Source," *Electronic Design*, October 15, 1981.

Wurzburg, Henry. "Regulator Performs Symmetry Correction in a Push-Pull Switching Power-Supply," *Electronic Design*, September 13, 1978, p. 142.

Wyatt, Michael A. "Generate Multiple Low-Level DC Voltages with a Two-Chip DC-DC Converter," *Electronic Design*, January 8, 1981, p. 272.

VMOS POWER TRANSISTORS

Barlage, Michael F. "Exploit VMOS FET's Advantages to Drive Bipolar Power Transistors," *EDN*, November 5, 1978, p. 93.

Frey, Gary D. "VMOS Power Amplifiers," *EDN*, November 5, 1977, p. 83.

Hamstra, Robert H. "Bipolar and VMOS Hybrid Delivers Fast Power Pulses," *Electronics*, December 6, 1979, p. 133.

Harvey, Barry. "Power-FET Amplifier Designs Boost Fidelity, Reduce Complexity," *EDN*, September 20, 1980, p. 137.

Hower, Philip L. "Bipolar vs MOSFET," *Electronics*, December 18, 1980, p. 106.

Jung, Walter. *IC Array Cookbook*, New York: Hayden Book Company, Inc. 1980, Chapter 4.

Krausse, Jurgen Tihanyi and J. Tillmanns, P. "Power MOSFET's Run Directly Off TTL," *Electronics*, August 28, 1980, p. 145.

Mattera, Lucinda. "Current, Voltage Ratings Climb for Power MOSFETs," *Electronic Design*, December 6, 1978, p. 28.

Oxner, Ed. "Will VMOS Power Transistors Replace Bipolars in HF Systems?" *EDN*, June 20, 1977, p. 71.

Roehr, Bill. "VMOS A Giant Step Toward the Ideal," *Radio Electronics*, July 1980, p. 55.

Severns, Rudy. "MOSFETs Rise to New Levels of Power," *Electronics* May 22, 1980, p. 143.

Shaeffer, Lee. "Use VMOS to Switch High Currents," *Electronic Design*, April 26, 1976, p. 66.

———. "VMOS Peripheral Drivers Solve High-Power Load Interface Problems," *Computer Design*, December, 1977, p. 90.

Stone, Robert S. and Berlin, Howard M. *Design of VMOS Circuits, with Experiments*. Indianapolis: Howard W. Sams & Co. 1980.

Twaddel, William. "ICs and Semiconductors," *EDN*, December 15, 1980, p. 116.

———. "Special Report on Semiconductors," *EDN*, October 5, 1979, p. 85.

Vander Kooi, Marvin and Ragle, Larry. "MOS Moves into High-Power Applications," *Electronics*, June 24, 1976, p. 98.

———. Marvin. "Replacing Bipolar Darlingtons with FETs Simplifies System Interfacing," *EET*, April 26, 1976, p. 36.

VOLTAGE/POWER MONITORING CIRCUITS

Albing, Bradley, "Comparator Detects Power-Supply Overvoltages, Catches Glitches," *Electronic Design*, July 19, 1979, p. 120.

Barrie, Gilbert. "Accurate Voltage Monitor Requires One IC," *EDN*, August 5, 1980, p. 125.

Bozian, Michael. "Find Out There's a "Brownout" Before Equipment Goes Haywire," *Electronic Design*, September 27, 1978, p. 98.

———. "Inexpensive Circuit Monitors Voltage Conditions in Battery-Operated Equipment," *Electronic Design*, March 1, 1978, p. 72.

Breeze, Eric G. and Cole, Earl V. "Opto-Isolated Monitor Provides Fail-Safe Control," *Electronics*, October 9, 1980, p. 162.

Burke, Marvin. "Chip Changes the Colors of Light-Emitting Diodes," *Electronics*, April 7, 1981, p. 158.

Crask, G.J. "Short Beep Means a Good Battery at Turn-On," *Electronic Design*, August 16, 1980, p. 189.

Kuljian, Maynard J. "Battery Monitor Operates on Only a Few Microamps," *Electronic Design*, October 25, 1976, p. 188.

Laabs, John. "Two-Chip Thermal Sensor/Switch Provides Control, Warning, and Shutdown," *Electronic Design*, September 27, 1980, p. 150.

Lefferts, Peter. "Power-Alarm Operates a Long Time on a Single 1.5 V Cell," *Electronic Design*, November 8, 1975, p. 108.

Okolowicz, John. "Automatic Charging Regulator Gives Overvoltage and Undervoltage Warnings," *Electronic Design*, March 15, 1976, p. 104.

Paluka, John R. "Overload Protector/Fault-Indicator Circuit," *NASA Tech Briefs*, Fall, 1976, p. 331.

Pease, Robert A. "Dithering' Display Expands Bar Graph's Resolution," *Electronics*, August 11, 1981, p. 114.

Rubinstein, Michael. "Low-Battery Threshold Detector Draws Only 2 μA Standby Current," *Electronic Design*, February 15, 1980, p. 156.

Rummel, Steve. "Reset Circuit Detects Power Drop-Out," *EDN*, May 20, 1976, p. 94.

Venditti, Carlo. "Protection Circuit Prevents IC Undervoltage," *EDN*, August 5, 1976, p. 80.

Venkatasubbiah, D. "Overvoltage/Undervoltage Indicator is Powered by the Circuit it Monitors," *Electronic Design*, May 24, 1980, p. 117.

Widlar, Robert; Dobkin, Robert; and Yamatake, Mineo. "Low-Voltage Op Amp Breakthrough Expands Linear Design Horizons," *EDN*, February 5, 1979, p. 97.

WINDOW COMPARATORS/DETECTORS

Chetty, P. R. K. "Simple Window Detector Handles 50 V," *EDN*, August 20, 1978, p. 152.

Graeme, Jerald. "Comparator Has Precise, Voltage-Controlled Hysteresis," *EDN*, August 20, 1975, p. 78.

Swartz, Robert. "Single Zener Sets Threshold for Simple Window Comparator," *Electronic Design*, August 16, 1978, p. 138.

Viles, R. S. "Upper and Lower Thresholds Can Be Set Independently in Latching-Comparator Circuit," *Electronic Design*, December 20, 1977, p. 176

Wallace, John D. "Norton Op Amp Teams with Zener for Single-Supply Window Comparator," *Electronic Design*, July 5, 1979, p. 86.

POWER SAVING DESIGNS/TECHNIQUES OR DEVICES

Chetty, P.R.K. "Logic-Function Generator Needs No Power Supply," *Electronics*, July 19, 1979, p. 128.

———. "DC-DC Converter Maintains High Efficiency," *Electronics*, January 13, 1981, p. 174.

Cole, H.A. "Display-Blanking Circuit Saves Power, Improves

Readability," *Electronic Design*, October 25, 1976, p. 188.

Colotti, James. "Load Power Sources for Peak Efficiency," *EDN*, October 5, 1979, p. 154.

Cowett, Philip M. "Switching Supply Converts −60 V to +5 and ±6.3 V With 83% Efficiency," *Electronic Design*, January 18, 1978, p. 106.

Edrington, Jim. "Power-Sharing Bridge Circuit Improves Amplifier Efficiency," *Electronics*, October 6, 1981, p. 142.

Ernst, Peter A. "Pulsing a PROM's Supply Voltage Greatly Reduces the Energy Used," *Electronic Design*, August 16, 1977, p. 102.

Kranz, Paul. "Simple Circuit Blanks the Leading Zeros for Single-Chip DVM/DPM A/D Converters," *Electronic Design*, September 22, 1977, p. 106.

Kularatna, A.D.V.N. "Optosensor Limits Shunt Supply's No-Load Current," *Electronics*, January 13, 1981, p. 174.

Kulturel, Sadeddin. "Switching Converter Raises Linear Regulator's Efficiency," *Electronics*, June 21, 1979, p. 126.

Lange, Jim. "Tapped Inductor Improves Efficiency for Switching Regulator," *Electronic Design*, August 2, 1978, p. 114.

Lefferts, Peter. "On Indicator Saves Power," *EDN*, April 20, 1979, p. 223.

———. "Power-Miser Flasher IC Has Many Novel Applications," *EDN*, March 20, 1976, p. 59.

Leiner, Jerome. "Demand-Switched Supply Boosts Amplifier Efficiency," *EDN*. November 9, 1978, p. 114.

Masters, George W. "Demand Power Supply Draws Low Standby Current," *Electronic Design*, September 13, 1977, p. 118.

Millard, G.J. "Solar-Powered Regulator Charges Batteries Efficiently," *Electronics*, September 13, 1979, p. 130.

Nussbaum, Stephen H. "Current Sources for I²L Saves Energy," *Electronics*, August 16, 1979, p. 115.

Pease, Robert A. "Bar-Graph Chip Lights LEDs in Series; Minimizes Current Drain for Displays," *Electronic Design*, January 18, 1980, p. 116.

Ranjan, M. Karunakar. "Simple Fix Cuts Display-Driver Dissipation," *EDN*, April 5, 1980, p. 137.

Shih, Kelvin. "Micropower Regulator Has Low Dropout Voltage," *Electronics*, April 12, 1979, p. 130.

Tang, Tee G. "Discrete Regulator Design Regulates Close to Supply Bus," *Electronic Design*, August 2, 1980, p. 102.

Trace de Almeida, Anibal. "Solar Power-Supply Regulator Consumes Less than 1 mA," *Electronic Design*, May 10, 1980, p. 244.

Venditti, Carlo. "Reducing the Dropout Voltage of Programmable Regulators," *Electronics*, October 11, 1979, p. 154.

HEATSINKING AND THERMODYNAMICS

Carbonell, Nelson P. and Nevala, David J. "Forced Air Cooling," *Electronic Products*, April, 1979, p. 59.

Chardon, Carlos C. "Design Equipment to Run Silent, Run Cool," *Electronic Design*, June 21, 1980, p. 119.

Martin, Vaughn D. "LCD Primer," *CQ*, April, 1983, p. 45.

Newberry, Deb. 'Heat Pipes," *Circuits Manufacturing*, June, 1983, p. 54.

Rodriguez, Edward T. "Model Semiconductor Thermal Designs Even with Scanty Vendor Data," *Electronic Design* February 15, 1979, p. 102.

Sass, Forrest. "Keep Cool and Live Longer," *Electronic Products*, May, 1980, p. 55.

Siegel, Bernard S. "Measuring Thermal Resistance is the Key to a Cool Semiconductor," *Electronics*, July 6, 1978, p. 121.

Index

Index

Page numbers set in *italics* refer to illustrations.